Surveys and Tutorials in the Applied Mathematical Sciences

Volume 12

Featuring short books of approximately 80-200pp, Surveys and Tutorials in the Applied Mathematical Sciences (STAMS) focuses on emerging topics, with an emphasis on emerging mathematical and computational techniques that are proving relevant in the physical, biological sciences and social sciences. STAMS also includes expository texts describing innovative applications or recent developments in more classical mathematical and computational methods.

This series is aimed at graduate students and researchers across the mathematical sciences. Contributions are intended to be accessible to a broad audience, featuring clear exposition, a lively tutorial style, and pointers to the literature for further study. In some cases a volume can serve as a preliminary version of a fuller and more comprehensive book.

Ferdinand Verhulst

A Toolbox of Averaging Theorems

Ordinary and Partial Differential Equations

Ferdinand Verhulst
Mathematical Institute
University of Utrecht
Utrecht, The Netherlands

ISSN 2199-4765 ISSN 2199-4773 (electronic)
Surveys and Tutorials in the Applied Mathematical Sciences
ISBN 978-3-031-34514-2 ISBN 978-3-031-34515-9 (eBook)
https://doi.org/10.1007/978-3-031-34515-9

This Springer imprint is published by the registered company Springer Nature Switzerland AG
The registered company address is: Gewerbestrasse 11, 6330 Cham, Switzerland

Preface

After spending many years of research on averaging related theory and applications, I was inclined to consider the theory of averaging for ODEs and PDEs as generally known. However, as became clear from many recent questions and comments, also from the persistent use of formal methods in the literature, this perception is not correct. It motivated me to write this toolbox of averaging methods as a summary of the theory with many theorems and examples; references to the proofs have been added. Simple and also more complicated applications illustrate the theory.

This part of my work was quite a long-lasting adventure. After studying theoretical astrophysics in Amsterdam where my graduate thesis was mainly in computing, I went in 1966 to the Technical University of Delft to do research for a PhD in applied mathematics. My thesis advisor, Wiktor Eckhaus, introduced me among many other things to averaging. The book by Bogoliubov and Mitropolsky [13] with its many ideas and the sharp text of Maurice Roseau [54] were a great help. With Adriaan van der Burgh and Jan Besjes, we modified and extended proofs of averaging. It was exciting to see that mechanical engineering in Delft used the Van der Pol-method, a kind of averaging, to study railway mechanics.

Much later, in 1969 and 1975, I met Yuri Mitropolsky in Kyev. When I noted that Krylov and Bogoliubov invented the multiple timescale method in the 1930s, he explained to me that they had concluded that the method would not be useful.

I moved to the University of Utrecht in 1971. Later, I had daily discussions there with Jan Sanders on averaging aspects and its relation with normalisation. This led to the first edition of our joint book on averaging for ODEs (*Averaging Methods for Nonlinear Dynamical Systems*, Springer, 1985). Together with an enthusiastic Jim Murdock, we produced a second, revised edition in 2007 [58], double in size but maybe less introductory.

Many discussions on qualitative aspects took place with Hans Duistermaat (if you find a periodic solution or a torus by averaging, does it really exist?), also with Richard Cushman (stimulating disagreement), Phil Holmes, Richard Rand and Taoufik Bakri. The interest of mathematical physicists in averaging and normalisation analysis in Italy, Milano (Giuseppe Gaeta) and Rome was a stimulus.

This also holds for the interest in engineering applications in Delft, Prague, Vienna, to name a few places often visited.

There seems to be no end to the extension and applicability of averaging methods.

Utrecht, The Netherlands   Ferdinand Verhulst
2023

Contents

Chapter 1
Introduction

The purpose of this book is to present a concise survey of averaging theorems as a toolbox for applied mathematicians, and especially also physicists and engineers. Most of the contents can be found in various parts of the scientific literature, we aim at a summary for practical use. New theory and new applications can be found in Chaps. 6–9.

As the emphasis is on the application of averaging we generally leave out proofs but refer to the references in the literature where proofs and validation can be found. Proofs are necessary but including them would double the size of the book; however, if we want to establish the validity of averaging results we need proofs, also if we find it necessary to extend the theory. That is why we add some discussion on the theorems.

In addition to the theorems we give elementary and more complicated examples to illustrate the theoretical results. We suppose that the reader is acquainted with basic (bachelor) knowledge of calculus, ordinary and partial differential equations (ODEs and PDEs). Apart from being a toolbox the book can also be useful for a mathematics seminar on averaging as the references to proofs of the theorems can be used to analyse the theoretical background.

In this first chapter we discuss possible approaches to perturbation problems. The formulation of variational equations for averaging is useful but the detailed form is depending on the type of problem. The chapter ends with a comparison of various perturbation theories and points out a hybrid strategy using both numerical and analytical approaches.

1.1 Perturbation Problems

We start quite general. Suppose we have a perturbation problem of the form

$$\dot{x} = f_0(t, x) + \varepsilon f_1(t, x) + \varepsilon^2 \ldots \tag{1.1}$$

© The Author(s), under exclusive license to Springer Nature Switzerland AG 2023
F. Verhulst, *A Toolbox of Averaging Theorems*, Surveys and Tutorials in the Applied Mathematical Sciences 12, https://doi.org/10.1007/978-3-031-34515-9_1

with x, f_0, f_1 n-vectors and suitable initial conditions. The parameter ε is small and positive. A dot like in $\dot{x}$ will always describe differentiation with respect to time. System (1.1) describes a perturbation of the ODE

$$\dot{y} = f_0(t, y). \qquad (1.2)$$

In later chapters we will see other perturbation formulations. We replaced x by y to indicate the quantitative and possibly qualitative difference between the perturbed (1.1) and unperturbed problem (1.2). The analysis of perturbation problems is subtle and full of surprises. To motivate our approach to develop variational equations in the subsequent sections we give 2 examples of a natural but deficient approach.

Example 1.1 A simple example that we can solve explicitly are damped harmonic oscillations described by the scalar initial value problem:

$$\ddot{x} + \varepsilon\dot{x} + x = 0, x(0) = 1, \dot{x}(0) = 0. \qquad (1.3)$$

Equations with higher order derivatives can be put in the form (1.1), for instance in this case by putting $x = x_1, \dot{x} = x_2$ and $x = (x_1, x_2)$. The unperturbed equation is the harmonic equation with goniometric functions as solutions. The perturbed equation produces quantitative and qualitative different solutions as the oscillations are slowly damped and tend to zero.

Ignoring the known exact solution one could quite naturally try to find an approximation by proposing the expansion:

$$x(t) = y_0(t) + \varepsilon y_1(t) + \varepsilon^2 \dots$$

Substitution in Eq. (1.3) produces:

$$\ddot{y}_0 + \varepsilon\ddot{y}_1 + \varepsilon(\dot{y}_0 + \varepsilon\dot{y}_1) + y_0 + \varepsilon y_1 + \dots = 0.$$

Collecting terms with equal powers of ε we find to first order:

$$\ddot{y}_0 + y_0 = 0, \ddot{y}_1 + \dot{y}_0 + y_1 = 0.$$

Applying the initial conditions to y_0 we find:

$$y_0(t) = \cos t, \ddot{y}_1 + y_1 = \sin t.$$

As we know, the equation for y_1 yields resonant solutions that increase with time and become unbounded, such terms are called *secular* terms. The terminology derives from eighteenth century astronomy. It is clear that we have to look for a different perturbation approach instead of a straightforward expansion. This will lead us to variational equations associated with system (1.1).

The next example deals with a classical equation of mathematical physics and engineering, the Mathieu-equation.

Example 1.2 If there are periodic terms in an equation, can we not immediately average without using complicated transformations? It would be attractive to apply such a procedure to ODEs with periodic terms. We consider the Mathieu-equation, a simple looking example but this is deceptive. Consider an oscillator with a small modulation of the frequency:

$$\ddot{x} + (1 + \varepsilon \cos 2t)x = 0, \, x(0) = 1, \, \dot{x}(0) = 0. \tag{1.4}$$

We will use this equation several times in the sequel. Solving the equation requires special functions and is not easy, but if $\varepsilon = 0$ the Eq. (1.4) is the harmonic equation. Can we use the harmonic solutions as a first approximation? An additional argument to use these solutions as an approximation for ε small is that averaging over time of the coefficient $\cos 2t$ gives zero. Applying the initial conditions we find for $\varepsilon = 0$:

$$x_0(t) = \cos t.$$

We expect $x_0(t)$ to be an approximation of the initial value problem (1.4). However, a numerical approximation of the initial value problem shows in Fig. 1.1 that the solution $x(t)$ is rather quickly increasing in size. The solution is in fact unbounded; we will discuss Mathieu-equations again later on.

Our approach in Example 1.2 seemed natural and it is surprising that this "crude averaging" gives erroneous results. Already in the eighteenth century mathematicians and astronomers realised this. It will turn out that the Lagrange "method of

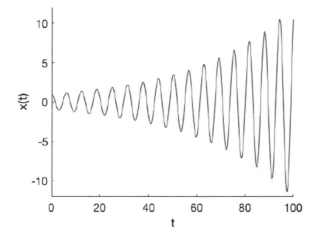

Fig. 1.1 Numerical approximation of the initial value problem (1.4)

variation of constants" leads to slowly varying variational equations that are suitable for averaging.

In physics and engineering it is quite natural to consider differential equations of the form (1.1) (or other equation types) that contain terms that are periodic in time t or periodic in a spatial variable. Such studies started in the eighteenth century when one considered gravitational two-body systems as the Sun and the Earth while one wanted to describe the motion of a smaller object in this system like the Moon or a comet. The idea is then to use the known solutions of the problem of two masses to study by variation of constants the problem with the perturbing small mass. Variation of constants is here in general a nonlinear transformation leading to *variational equations* that contain periodic terms which can be averaged.

1.2   Amplitude-Phase Transformation

Consider a second order equation of the form:

$$\ddot{x} + \omega^2 x = \varepsilon f(t, x, \dot{x}). \tag{1.5}$$

The frequency ω is a positive constant. We can solve the equation if $\varepsilon = 0$ with solutions of the form:

$$x(t) = r_0 \cos(\omega t + \phi_0).$$

The constant amplitude r_0 and phase ϕ_0 are determined by the initial conditions. We propose to apply the Lagrange method of *variation of constants* by transforming $x, \dot{x} \mapsto r, \phi$:

$$x = r(t)\cos(\omega t + \phi(t)), \quad \dot{x} = -r(t)\omega \sin(\omega t + \phi(t)). \tag{1.6}$$

Requiring that the derivative of x equals $\dot{x}$ and substituting these expressions in Eq. (1.5) we find:

$$\begin{cases} \dot{r}\cos(\omega t + \phi) - r\sin(\omega t + \phi)\dot{\phi} = 0, \\ -\dot{r}\omega\sin(\omega t + \phi) - r\omega\cos(\omega t + \phi)\dot{\phi}) = \varepsilon f(.), \end{cases} \tag{1.7}$$

where we abbreviated $f(.) = f(t, r(t)\cos(\omega t + \phi)(t), -r(t)\omega\sin(t + \phi(t))$. Solving algebraically for $\dot{r}, \dot{\phi}$ we find the so-called variational equations:

$$\begin{cases} \dot{r} &= -\frac{\varepsilon}{\omega}\sin(t + \phi)f(.), \\ \dot{\phi} &= -\frac{\varepsilon}{\omega r}\cos(t + \phi)f(.). \end{cases} \tag{1.8}$$

System (1.8) suggests that the variables amplitude and phase $r(t), \phi(t)$ are slowly changing with time. The solutions of system (1.8) give us with transformation (1.6) the exact solutions of Eq. (1.5). In general we cannot solve system (1.8) analytically but we can average the system over time keeping r, ϕ constant during the averaging process. In Chap. 2 we will give a precise formulation of the process and refer to proofs.

The variation of constants method leaves us the freedom to choose other independent solutions of the unperturbed system instead of the amplitude-phase representation, for instance a combination of $\cos t$ and $\sin t$. It turns out that for autonomous equations, if $f(t, x, \dot{x})$ does not explicitly depends on t, the amplitude phase transformation (1.6) is quite effective.

We demonstrate the use of variational equations by some examples.

Example 1.3 We consider an oscillator with small damping $\varepsilon \dot{x}$ and small anharmonic term $-\varepsilon x^3$:

$$\ddot{x} + \varepsilon \dot{x} + x - \varepsilon x^3 = 0, \tag{1.9}$$

with initial values $x(0), \dot{x}(0)$ given. The equation is autonomous. Comparing with Eq. (1.5) we have $\omega = 1$ and $f = -\varepsilon \dot{x} + \varepsilon x^3$.

Introducing amplitude-phase transformation (1.6) we find with (1.9) the variational system:

$$\begin{cases} \dot{r} &= -\varepsilon \sin(t + \phi)\left(r \sin(t + \phi) + r^3 \cos^3(t + \phi)\right), \\ \dot{\phi} &= -\frac{\varepsilon}{r} \cos(t + \phi)\left(r \sin(t + \phi) + r^3 \cos^3(t + \phi)\right). \end{cases} \tag{1.10}$$

The righthand side of the system is 2π-periodic in t. Anticipating the formulation in Chap. 2 we find the averages:

$$\frac{1}{2\pi} \int_0^{2\pi} \sin(t + \phi)\left(r \sin(t + \phi) + r^3 \cos^3(t + \phi)\right) dt = \frac{1}{2}r,$$

$$\frac{1}{2\pi} \int_0^{2\pi} \cos(t + \phi)\frac{1}{r}\left(r \sin(t + \phi) + r^3 \cos^3(t + \phi)\right) dt = \frac{3}{8}r^2.$$

The averaged variational system becomes:

$$\frac{dr}{dt} = -\frac{\varepsilon}{2}r, \quad \frac{d\phi}{dt} = -\varepsilon\frac{3}{8}r^2. \tag{1.11}$$

We kept the notation r, ϕ although the averaged system is supposed to produce an approximation. The approximations are with initial values $r(0), \phi(0) = 0$:

$$r(t) = r(0)e^{-\frac{\varepsilon}{2}t}, \, \phi(t) = \frac{3}{8}e^{-\varepsilon t} - \frac{3}{8}.$$

As a second example we consider a more general anharmonic equation.

Example 1.4 Consider the anharmonic oscillator described by the nonlinear equation:

$$\ddot{x} + x - \varepsilon x^n = 0, \tag{1.12}$$

with initial values $x(0), \dot{x}(0)$ given, n is a natural number $n = 2, 3, 4, \ldots$ As in the preceding example we transform $x, \dot{x} \mapsto r, \phi$ by Eq. (1.6) and apply again the Lagrange method of *variation of constants*.

For $\dot{r}, \dot{\phi}$ we find from system (1.8) the variational equations:

$$\begin{cases} \dot{r} & = -\varepsilon r^n \sin(t + \phi) \cos^n(t + \phi), \\ \dot{\phi} & = -\varepsilon r^n \cos^{n+1}(t + \phi). \end{cases} \tag{1.13}$$

Averaging the righthand sides over t keeping r and ϕ constant we find:

$$\frac{d\tilde{r}}{dt} = 0, \quad \frac{d\tilde{\phi}}{dt} = 0 \text{ if n is even}, \tag{1.14}$$

whereas $d\tilde{\phi}/dt \neq 0$ if $n =$ odd. For instance if $n = 3$ we find:

$$\frac{d\tilde{\phi}}{dt} = -\varepsilon \frac{3}{8} r(0)^2.$$

The result for the approximate amplitude $\tilde{r}$ should not surprise us. As we can verify by differentiation equation (1.12) has the first integral

$$\frac{1}{2}\dot{x}^2 + \frac{1}{2}x^2 - \varepsilon \frac{1}{n+1} x^{n+1} = \text{constant}, \tag{1.15}$$

corresponding with closed curves, cycles, in the phase-plane. The constant is determined by the initial conditions. The cycles are ε-close to the circles $\dot{x}^2 + x^2 = \text{constant}$. A question is then whether a more accurate approximation might produce non-trivial results for the amplitude; such questions are studied in Chap. 4. On the other hand, the phase changes non-trivially if n is odd, corresponding with an $O(\varepsilon)$ change of the period 2π depending on $r(0)$.

1.3   Comoving Varables

Consider again the perturbed harmonic equation (1.5). Using amplitude-phase transformtion (1.6) may give problems when considering small values of the amplitude as polar coordinates become singular near $r = 0$. We can see this in the variational system (1.8) where r is found in the denominator of the phase-equation. However, if we can write after transformation to amplitude-phase coordinates:

$$\varepsilon f(.) = \varepsilon r(t) \bar{f}(t, r(t) \cos(\omega t + \phi)(t), -r(t)\omega \sin(t + \phi(t))$$

with $\bar{f}$ non-singular in r we have a removable singularity at $r = 0$ and we can still use amplitude-phase variables. This is for instance the case in Example 1.4.

If we have a singularity in the variational system for $r = 0$ we choose other independent solutions of the unperturbed system, in this case $x, \dot{x} \mapsto y_1, y_2$ with:

$$x = y_1 \cos \omega t + \frac{y_2}{\omega} \sin \omega t, \quad \dot{x} = -y_1 \omega \sin \omega t + y_2 \cos \omega t. \tag{1.16}$$

The variational equations become with the same procedure as before:

$$\begin{cases} \dot{y}_1 &= -\frac{\varepsilon}{\omega} \sin(\omega t) f(t, y_1 \cos \omega t + \frac{y_2}{\omega} \sin \omega t, -y_1 \omega \sin \omega t + y_2 \cos \omega t), \\ \dot{y}_2 &= \varepsilon \cos(\omega t) f(t, y_1 \cos \omega t + \frac{y_2}{\omega} \sin \omega t, -y_1 \omega \sin \omega t + y_2 \cos \omega t). \end{cases}$$

$$\tag{1.17}$$

It turns out that for nonautonomous perturbations transformation (1.16) has advantages, for autonomous problems transformation (1.6) is useful as it gives directly equations for the change of amplitude and phase. We will see many examples in later chapters, but we discuss an iconic example already here.

The Mathieu-Equation with Prime Resonance

Example 1.5 Consider again the so-called Mathieu-equation from Example 1.2:

$$\ddot{x} + (1 + 2\varepsilon \cos 2t)x = 0.$$

Because of the presence of the term $2\varepsilon \cos 2t \, x$ this linear equation is called *parametrically excited*. In slowly-varying phase-amplitude form, using the transformation (1.6) we find the variational system:

$$\begin{cases} \dot{r} &= 2\varepsilon \sin(t + \phi) \cos 2t \, r \cos(t + \phi), \\ \dot{\psi} &= 2\varepsilon \cos^2(t + \phi) \cos 2t. \end{cases} \tag{1.18}$$

Averaging over the 2π-periodic righthand side keeping r, ϕ constant produces

$$\dot{r} = \frac{1}{2}\varepsilon r \sin 2\phi, \quad \dot{\phi} = \frac{1}{2}\varepsilon \cos 2\phi.$$

This system is not convenient to analyse, it is even nonlinear. Also we would like to study the behaviour near the origin where $r = 0$.

Trying the second transformation (1.16) we obtain the variational system:

$$\begin{cases} \dot{y}_1 &= 2\varepsilon \sin t \cos 2t \, (y_1 \cos t + y_2 \sin t), \\ \dot{y}_2 &= -2\varepsilon \cos t \cos 2t \, (y_1 \cos t + y_2 \sin t). \end{cases} \tag{1.19}$$

Averaging the 2π-periodic system keeping y_1, y_2 constant produces:

$$\dot{y}_1 = -\frac{1}{2}\varepsilon y_2, \quad \dot{y}_2 = -\frac{1}{2}\varepsilon y_1. \tag{1.20}$$

These are linear equations with constant coefficients and are easier to analyse; the eigenvalues near $(0, 0)$ are $\pm\frac{1}{2}\varepsilon$ so that the approximations of y_1, y_2 are a linear combination of $\exp(\pm\frac{1}{2}\varepsilon t)$. If for instance $x(0) = x_0$, $\dot{x}(0) = 0$, we have $y_1(0) = x_0$, $y_2(0) = 0$ and

$$y_1(t) = \frac{1}{2}x_0 e^{-\frac{1}{2}\varepsilon t} + \frac{1}{2}x_0 e^{\frac{1}{2}\varepsilon t}, \quad y_2(t) = \frac{1}{2}x_0 e^{-\frac{1}{2}\varepsilon t} - \frac{1}{2}x_0 e^{\frac{1}{2}\varepsilon t},$$

and for $x(t)$:

$$x(t) = \frac{1}{2}x_0 e^{-\frac{1}{2}\varepsilon t}(\cos t + \sin t) + \frac{1}{2}x_0 e^{\frac{1}{2}\varepsilon t}(\cos t - \sin t).$$

These approximate solutions of the Mathieu-equation show the instability of the trivial solution at, what is usually called, the prime resonance of the Mathieu-equation.

1.4   Amplitude-Angle Transformation

Consider again the perturbed harmonic equation (3.3). Instead of amplitude-phase transformation we can also transform to amplitude and angle $x, \dot{x} \mapsto r, \alpha$:

$$x = r \sin \alpha, \quad \dot{x} = \omega r \cos \alpha \tag{1.21}$$

The transformation (1.21) leads by variation of constants to:

$$\begin{cases} \dot{r} &= \frac{\varepsilon}{\omega} \cos \alpha f(t, r \sin \alpha, \omega r \cos \alpha), \\ \dot{\alpha} &= \omega - \frac{\varepsilon}{\omega r} \sin \alpha f(t, r \sin \alpha, \omega r \cos \alpha). \end{cases} \tag{1.22}$$

The angle variable α replaces $t + \phi$ of the amplitude-phase transformation, α will be *timelike*, the phase of the oscillations is also a type of angle but not necessarily timelike. If we have more, say n coupled oscillators, we can introduce in the same way n amplitudes and n angles. We shall see applications with amplitude-angle formulations in Chap. 7.

For other special equations from mathematical physics, for instance Hamiltonian systems, other variational equations may be appropriate; see Chap. 8.

1.5   Perturbed Linear Systems and Forcing

When studying perturbation problems it is natural to use our knowledge of the unperturbed ($\varepsilon = 0$) problem but as we have seen there are various ways to formulate the solutions of the 'unperturbed' problems. To obtain variational (slowly varying) equations is easiest in the case that the unperturbed problem is linear, and especially if the coefficients are constant. Consider in $\mathbb{R}^n$ the quasi-linear system with continuous $n \times n$ matrix $A(t)$ and perturbation term $\varepsilon g(t, x) + \varepsilon^2 \ldots$:

$$\frac{dx}{dt} = A(t)x + \varepsilon g(t, x) + \varepsilon^2 \ldots$$

The vector function $g \in \mathbb{R}^n$ and the dots represent functions that are smooth (continuously differentiable) in x, continuous in t on suitable domains (to be specified later). The unperturbed problem is linear and has n independent solutions that we group together in a fundamental matrix $\Phi(t)$. In many problems $A(t)$ is constant and this is the easiest case.

The variation of constants method of Lagrange suggests the transformation $x \rightarrow z$:

$$x = \Phi(t)z$$

producing the *variational equation*:

$$\frac{dz}{dt} = \varepsilon \Phi^{-1}(t)g(t, \Phi(t)z) + \ldots$$

We expect $z(t)$ to be a slowly varying function of time. As we have seen, the choice of independent solutions may change the appearance of the variational equation, of course without changing the resulting solutions of the perturbed ODE. However, a good choice of independent solutions helps us to understand the calculations of the perturbation process.

If the unperturbed ($\varepsilon = 0$) system consists of a linear system with constant coefficients plus a forcing term we can exploit our knowledge of such problems as we often know the solutions of such forced linear systems explicitly.

Consider the system in vector form:

$$\dot{x} = A(t)x + \phi(t) + \varepsilon g(t, x), \tag{1.23}$$

with $x \in \mathbb{R}^n$, $A(t)$ a $n \times n$ continuous matrix, $\phi(t)$ a continuous n-vector, $g(t, x)$ is a sufficiently smooth n-vector. Suppose that $\Phi(t)$ is a fundamental matrix of the homogeneous equation $\dot{x} = A(t)x$, the n-vector $\eta(t)$ is a particular solution of the inhomogeneous equation if $\varepsilon = 0$. In general we will know $\Phi(t)$ and $\eta(t)$ explicitly only if $A(t)$ is constant. A general solution $x_l(t)$ of the unperturbed system is

$$x_l(t) = \Phi(t)c + \eta(t),$$

with c a constant n-vector. We transform $x \rightarrow y$ by:

$$x = \Phi(t)y + \eta(t). \tag{1.24}$$

Substitution in system (1.23) produces:

$$\Phi(t)\dot{y} = \varepsilon g(t, \Phi(t)y + \eta(t)).$$

By definition, a fundamental matrix is non-singular, its inverse exists. It follows that we have:

$$\dot{y} = \varepsilon \Phi(t)^{-1} g(t, \Phi(t)y + \eta(t)). \tag{1.25}$$

An initial condition $x(t_0)$ applied to system (1.23) will transform to:

$$y(t_0) = \Phi^{-1}(t_0)(x(t_0) - \eta(t_0)).$$

Example 1.6 In the case of the forced harmonic equation

$$\ddot{x} + \omega^2 x = \psi(t) + \varepsilon g(t, x, \dot{x}), \tag{1.26}$$

with nontrivial scalar solution $x(t) = \eta(t)$ if $\varepsilon = 0$ we transform:

$$x = r\cos(\omega t + \phi) + \eta(t), \ \dot{x} = -r\omega\sin(\omega t + \phi) + \dot{\eta}(t). \tag{1.27}$$

This leads to the slowly varying system obtained by transformation (1.27) in the form:

$$\begin{cases} \dot{r} & = -\frac{\varepsilon}{\omega}\sin(t + \phi)g(t, r\cos(\omega t + \phi) + \eta(t), -r\omega\sin(\omega t + \phi) + \dot{\eta}(t)), \\ \dot{\phi} & = -\frac{\varepsilon}{\omega r}\cos(t + \phi)g(t, r\cos(\omega t + \phi) + \eta(t), -r\omega\sin(\omega t + \phi) + \dot{\eta}(t)). \end{cases}$$
$$\tag{1.28}$$

In Chap. 6 this technique is applied to the forced Duffing-equation. The forcing complicates the dynamics considerably.

1.6   Order Functions and Timescales

To estimate accuracy and validity of approximations we have to introduce some concepts that may look boring but that are absolutely necessary to obtain precise statements. Consider for instance the function $f(t) = \sin t + \sqrt{\varepsilon} \cos t + \varepsilon^2 t$. The function $\sin t$ can be considered an approximation of $f(t)$ but in what sense? A guess is that it is an approximation with accuracy $\sqrt{\varepsilon}$ but is this true for all time? We will introduce the concepts for statements like this to make sense.

We will give a short description of norms and estimates for ODEs. The extension to PDEs is to a large extent analogous, see Chap. 10. In the sequel, for ODEs, the norm of a constant n-vector u is $\|u\| = \sum_{i=1}^{n} |u_i|$, the norm of an $n \times n$ matrix A is $\|A\| = \sum_{i,j=1}^{n} |a_{ij}|$. When considering finite-dimensional vector functions depending on variables and defined on a certain domain D, we will use the supremum norm on D.

Estimating a function $f(t, x, \varepsilon)$ dependent on time, n-dimensional variable $x \in D$ and parameter ε means that we take the supremum norm of $f(t, x, \varepsilon)$ on the chosen time interval and D resulting in an expression in ε. We have some abuse of language here as the norm of a vector function is supposed to be a real number instead of a function of parameter ε.

Suppose that we want to approximate a (vector) function $x_\varepsilon(t)$ for $t \geq 0$ by another function $y_\varepsilon(t)$ in terms of the small parameter ε. For the concept of 'approximation' we need so-called order-functions and timescales.

Definition 1.1 A real, continuous function $\delta(\varepsilon)$ on $(0, 1]$ for which $\lim_{\varepsilon \to 0}$ exists will be called an order function.

Examples of order functions are $1 + \varepsilon, \varepsilon, \varepsilon^2, \varepsilon \ln \varepsilon, \varepsilon^{2/5}$ etc.

An important concept is the timescale, where an approximation is valid. A timescale indicates the length of a time-interval measured in ε or better $1/\varepsilon$.

Definition 1.2 Consider the vector functions $x_\varepsilon(t)$ and $y_\varepsilon(t)$. If we have for $t \geq 0$ the estimate

$$x_\varepsilon(t) - y_\varepsilon(t) = O(\varepsilon^m), \quad 0 \leq \varepsilon^n t \leq C,$$

with m, n, C positive constants independent of ε, we shall say that $y_\varepsilon(t)$ is an $O(\varepsilon^m)$-approximation of $x_\varepsilon(t)$ on the timescale $1/\varepsilon^n$.

Note that if necessary we can generalise the formulation somewhat by replacing ε^m by the order function $\delta_1(\varepsilon)$ and ε^n by the order function $\delta_2(\varepsilon)$.

Example 1.7 Consider the function $x_\varepsilon(t) = (1 + \varepsilon + \varepsilon^2 t) \sin t$ with $O(\varepsilon)$-approximation $\sin t$ on the timescale $1/\varepsilon$. However, $\sin t$ is also an $O(\varepsilon^{\frac{1}{2}})$-approximation on the timescale $1/\varepsilon^{\frac{3}{2}}$. It is also easy to see that $(1 + \varepsilon) \sin t$ is an $O(\varepsilon^2)$-approximation on the timescale 1.

We shall often use this abbreviation of "an approximation on a timescale". Usually we will be happy to obtain an approximation on a timescale such as $1/\varepsilon$ or $1/\varepsilon^2$. As the estimates are valid for $\varepsilon \to 0$, these are already long timescales.

We started this section with the example $f(t) = \sin t + \sqrt{\varepsilon} \cos t + \varepsilon^2 t$ that can be handled in a similar way.

1.7   On Contraction and Iteration Approximations

A basic assumption for the existence and uniqueness of ODE initial value problems is to assume that the ODE ($\dot{x} = f(t, x)$) has a righthand side that is Lipschitz-continuous in x and continuous in t. A more precise formulation runs as follows. Consider the equation:

$$\dot{x} = f(t, x), \; x(t_0) = x_0,$$

with $x, x_0 \in D \subset \mathbb{R}^n, t, t_0 \in \mathbb{R}, f(t, x)$ is continuous in $[t_0, \infty)$. For $f(t, x)$ we have with positive constant k and x_1, x_2 in a neighbourhood of x_0 and for $t_0 \le t \le t_0 + L$ (L a positive constant) that in the Euclidean norm holds the estimate:

$$\|f(t, x_1) - f(t, x_2)\| \le k\|x_1 - x_2\|.$$

One can interpret the estimate as stating that locally the change of $f(t, x)$ is bounded by linearisation. Note that if the vector field $f(t, x)$ is continuously differentiable in a neigbourhood of x_0 it is Lipschitz-continuous. We have now a remarkable result, see [20] chs. 1 and 2.

For the initial value problem we formulate an equivalent Volterra integral equation:

$$x(t) = x_0 + \int_{t_0}^{t} f(s, x(s))ds. \tag{1.29}$$

If we have no idea of the form of the solution we may use as a very crude first approximation $x^0(t) = x_0$. The next approximations will be:

$$x^{(1)}(t) = x_0 + \int_{t_0}^{t} f(s, x_0)ds, \; x^{(2)}(t) = x_0 + \int_{t_0}^{t} f(s, x^{(1)}(s))ds, \; x^{(3)}(t) = \dots,$$

etc. The surprise is that even with this constant start function we have that after an infinite number of iteration steps the sequence of functions $x_0, x^{(1)}(t), x^{(2)}(t), \dots$ converges to the solution of the initial value problem. The iteration process is called *Picard-Lindelöf iteration* and in a more abstract setting *contraction*.

The process with this simple start function is important to obtain existence and uniqueness of solutions but not for actual constructions or approximations of the solution. It is instructive to apply the iteration process to the harmonic oscillator with initial conditions, for instance:

$$\ddot{x} + x = 0, \ x(0) = 0, \ \dot{x}(0) = 1,$$

with solution $\sin t$. We change it into a first order vector ODE by putting $x = x_1$, $\dot{x} = \dot{x}_1 = x_2$, $\dot{x}_2 = -x_1$, $x_1(0) = 0$, $x_2(0) = 1$. The integral equations become:

$$x_1(t) = \int_0^t x_2(s)ds, \ x_2(t) = 1 - \int_0^t x_1(s)ds.$$

Using as start function the initial conditions $0, 1$ we find at the second step $t, 1$, next $t, 1 - \frac{1}{2}t^2$, etc. The iteration process produces a series of polynomials in t that we recognise as the Taylor expansion of $\sin t$. But in general it will not be possible to identify the nature of the iteration series. Also, truncation of the series restricts the timescale of validity of the Taylor approximation severely.

However, the iteration process works equally well and sometimes better with a good choice of starting function. We will use this idea in Chaps. 3 and 8 for periodic solutions and quasi-periodic solutions. The general approach is as follows. Consider again Eq. (1.1) but for simplicity leave out the $O(\varepsilon^2)$ terms:

$$\dot{x} = f_0(t, x) + \varepsilon f_1(t, x).$$

Assume that we can solve the unperturbed problem with initial condition x_0, call this solution $x^{(0)}(t)$. Use this solution as a start of the iteration process, so

$$
\begin{aligned}
x^{(1)}(t) &= x^{(0)}(t) + \varepsilon \int_{t_0}^t f_1(s, x^{(0)}(s))ds, \\
x^{(2)}(t) &= x^{(1)}(t) + \varepsilon \int_{t_0}^t f_1(s, x^{(1)}(s))ds,
\end{aligned}
$$

$$\cdots \qquad \qquad \cdots$$

$$x^{(n)}(t) = x^{(n-1)}(t) + \varepsilon \int_{t_0}^t f_1(s, x^{(n-1)}(s))ds.$$

The process will converge for $n \to \infty$ by contraction but there is one problem. The iteration process is in general convergent for $t_0 \le t \le t_0 + L$ and so on a timescale $O(1)$ with respect to ε in a neighbourhood of the initial value in D. This makes the iteration process useful only for bounded solutions on an $O(1)$ interval of time, for instance for periodic solutions and quasi-periodic solutions if they exist. See Chaps. 3 and 9.

1.8   Comparison of Methods

This section contains general information on approximation methods and is independent of the sequel.

The reader will notice that in this book other analytic methods get scarce attention but as perturbation theory is an old subject, we should realise many different methods have been formulated in the literature. See [79] for a more extensive comparison with other methods and references.

Multiple Timing
In particular the method of multiple timescales or multiple timing is intuitively attractive, it has become a popular method. The idea of multiple timing is as follows. Suppose we have a perturbation problem like

$$\ddot{x} + x = \varepsilon f(t, x)$$

with x and f scalar functions, or more general a problem like (1.1). The solutions and its approximations are expected to be functions of t, εt, $\varepsilon^2 t$ or even $\varepsilon^n t$ with $n = 3, 4, \ldots$ Assume then that the approximation of $x(t)$ is a function of different variables t, $\tau_1 = \varepsilon t$, $\tau_2 = \varepsilon^2 t$ etc, so $x = x(t, \tau_1. \tau_2, \ldots)$ It is possible in this way to derive equations that produce a formal approximation of the solutions. However, the formal approximations obtained by the procedure are completely dependent on the apriori assumptions of certain timelike varables.

For elementary problems that can be treated by *first order averaging*, see Chap. 2, the averaging approximations are dependent on t and εt, they are equivalent to multiple timing calculations. *For second order calculations and more advanced research problems multiple timing gives sometimes erroneous results, the method depends on luck and correct guesses.* This is caused by the *apriori choice* of relevant timelike variables whereas in other methods, like averaging, the timelike variables emerge naturally during the calculations; for more discussion see [77] and [79] where this is explicitly shown for the Mathieu-equation.

The emergence of unexpected timelike variables is often connected with bifurcations. We present a simple example:

$$\dot{x} = \varepsilon^2 y, \ \dot{y} = -\varepsilon x, \ \ddot{z} + z = x, \tag{1.30}$$

with initial values $x(0) = 0$, $y(0) = 1$, $z(0) = \dot{z}(0) = 0$. Solving the system to $O(\varepsilon)$ and indicating the approximation by $\tilde{x}$ etc. we find:

$$\tilde{x} = 0, \ \tilde{y} = 1, \tilde{z} = 0.$$

We can easily solve the system with the initial values exactly to find the solutions:

$$x(t) = \sqrt{\varepsilon} \sin \varepsilon^{\frac{3}{2}} t, \ y(t) = \cos \varepsilon^{\frac{3}{2}} t, \ z(t) = \frac{\sqrt{\varepsilon}}{1 - \varepsilon^3} \sin \varepsilon^{\frac{3}{2}} t - \frac{\varepsilon^2}{1 - \varepsilon^3} \sin t.$$

So the (exact) solutions show unexpected timelike variables. See for more discussion on timelike variables Sect. 4.2.3.

Normalisation
The basic approach for normalisation is to transform the differential equation to a more tractable form. The resulting equations after transformation have usually to be truncated at the higher order terms, the truncated result is called a normal form. Consider for instance the system:

$$\dot{x} = Ax + f(x),$$

with x a n-vector, A a $n \times n$ constant matrix, $f(x) = f_2(x) + f_3(x) + \ldots + f_n(x)$ where $f_n(x)$ is a homogeneous polynomial of degree n. We introduce a certain resonance characteristic of the eigenvalues of matrix A; for details see [76, ch. 13] or [31]. If the eigenvalues of the matrix A are non-resonant, the equation can be put into linear form by a near-identity transformation $x = y + h_2(y) + h_3(y) + \ldots$ where the terms $h_j, \ j = 2, 3, \ldots$ are homogeneous polynomials in y. Poincaré showed that if the eigenvalues are non-resonant the equation for x can be transformed to the linear equation $\dot{y} = Ay$.

In [31, section 3.3], an explicit calculation is given for he perturbed harmonic oscillator like the Van der Pol-equation. In this case $n = 2$ and the eigenvalues of A are $\lambda_{1,2} = \pm i$. There is resonance but all quadratic terms can still be removed.

The names of Poincaré and Dulac are connected with the first formulation of normalisation. Near-identity transformations are related to the analysis that takes place to produce averaging approximations. Details about this relation are given in [58], see also remark 2.1 in Chap. 2.

The Poincaré-Lindstedt Method
This method focuses on periodic solutions of ODEs. Lindstedt developed the method for the Eq. (1.5) and two coupled equations of this type; Lindstedt's analysis is ingeneous but formal in the sense that it produces no existence results and no approximation estimates. Poincaré extended the method for general ODEs and introduced as basic principle the use of continuation based on the implicit function theorem and contraction. He concluded to existence of the periodic solutions and convergence of the expansions. See also Sect. 1.7 and for examples Chap. 3 and references [76]. There is a relation with averaging as one needs expansion of amplitudes and phases (or other variables) and one eliminates secular terms by imposing conditions. We will discuss the method again in Chap. 3.

Renormalisation
This is a formal method (no existence and no approximation estimates) that has some elements of the Poincaré-Lindstedt method but considers more general problems and solutions. The method avoids making assumptions on timelike variables and so obtains better results than multiple timing. For references and more details zee [79].

The Method of Harmonic Balance
We follow the description in [63]. Looking for periodic solutions of an ODE in x one can substitute a Fourier expansion of the form:

$$x(t) = \sum_{n=0}^{\infty} (a_n \cos n\Omega t + b_n \sin n\Omega t).$$

The parameter Ω is still unknown. The idea is then to equate after substitution the coefficients of equal harmonics. A simple example is to apply this for the equation:

$$\ddot{x} + x = cx^3, \tag{1.31}$$

with parameter c. Usually one chooses a finite expansion, for instance $n = 0, 1, a_0 = 0$; we find after substitution of the Fourier expression $a_1 \cos \Omega t + b_1 \sin \Omega t$:

$$-a_1 \Omega^2 \cos \Omega t - b_1 \Omega^2 \sin \Omega t + a_1 \cos \Omega t + b_1 \sin \Omega t = c(a_1 \cos \Omega t + b_1 \sin \Omega t)^3.$$

We have the conditions:

$$-a_1 \Omega^2 + a_1 = ca_1^3 \frac{3}{4}, \quad -b_1 \Omega^2 + b_1 = cb_1^3 \frac{3}{4},$$

where we have neglected higher order harmonics. The parameters Ω and a_1, b_1 depend on each other. This makes sense as we know that Eq. (1.31) is conservative and has a familiy of periodic solutions with period depending on the amplitude. However, we have in general no clear insight in the existence and approximation character of our results.

Conclusion
Apart from explicit error estimates, an important aspect of approximation results is the question whether a formally approximate result like a periodic solution or a quasi-periodic family of solutions corresponds with the same phenomena in the original unperturbed system. Only when using averaging methods and the Poincaré-Lindstedt method we will have both qualitative and approximation statements, see for instance Chap. 3 (periodic solutions) and Chap. 9 (quasi- or almost-periodic solutions).

1.9 Analytic and Numerical Approximations

An interesting question is why, with the advance of numerical computing, one would bother with calculating analytical approximations. The answer is that one should realise that in real-life problems of engineering, physics, biology and other sciences we have many parameters, much more than in the examples of this chapter, and also more variables leading to spatial dimensions much higher than two. With so many choices to make without advance knowledge of the solutions numerical calculations look arbitrary. How do we know to make a choice? Numerically one obtains illustration and confirmation but not validation of global results.

Considering differential equations with several parameters and also initial or boundary conditions a sensible research strategy is combining analytic approximation theory with high accuracy numerics. Useful numerical routines can be found for instance in the programs AUTO, MATCONT and MATHEMATICA; we will show more details and examples later. The analytic approximation computations point at dependence on parameters and side conditions, it also supplies more general qualitative insight. The numerics adds experimentation with inspiring phenomena, it adds necessary special illustrations and confirmation.

Examples of such a hybrid approach can be found in [8] and [6].

Chapter 2
First Order Periodic Averaging

We have seen averaging results in the Introduction. We will explain now the theory for $O(\varepsilon)$ approximations on a long timescale. Consider the variational equation with initial value:

$$\dot{x} = \varepsilon f(t, x) + \varepsilon^2 g(t, x, \varepsilon), \ x(t_0) = x_0. \tag{2.1}$$

Suppose that $x \in \mathbb{R}^n$ and $f(t, x)$ is T-periodic with respect to time t, T is a positive constant independent of ε; we average the vector function f with respect to time keeping x fixed:

$$f^0(x) = \frac{1}{T} \int_{t_0}^{t_0+T} f(t, x) dt.$$

We will use the notation $Df(t, x)$ or for short Df for the Jacobian. If we have the n-dimensional vector function $f(t, x) = (f_1, f_2, \ldots, f_n)^t$, then $Df(t, x)$ is the $n \times n$ matrix where we differentiate the elements only to the spatial variable $x = (x_1, x_2, \ldots, x_n)$:

$$Df = \begin{pmatrix} \partial f_1/\partial x_1 & \partial f_1/\partial x_2 & \ldots & \partial f_1/\partial x_n \\ \partial f_2/\partial x_1 & \partial f_2/\partial x_2 & \ldots & \partial f_2/\partial x_n \\ \ldots & \ldots & \ldots & \ldots \\ \partial f_n/\partial x_1 & \partial f_n/\partial x_2 & \ldots & \partial f_n/\partial x_n \end{pmatrix}. \tag{2.2}$$

2.1 The Basic Averaging Theorem

We have the following basic theorem for periodic averaging:

Theorem 2.1 *Consider the initial value problem* (2.1) *and the initial value problem*

$$\dot{y} = \varepsilon f^0(y), \ y(t_0) = x_0. \tag{2.3}$$

Assume for $x, y, x_0 \in D \subset \mathbb{R}^n, t_0 \le t \le \infty, 0 \le \varepsilon \ll 1$:

1. *f, g, Df are defined, continuous and bounded by a constant M independent of ε in $[t_0, \infty] \times D$;*
2. *g is Lipschitz-continuous with respect to $x \in D$;*
3. *f is T-periodic in T with T a constant independent of ε;*
4. *$y(t)$ belongs to a bounded interior subset of D on the timescale $1/\varepsilon$,*

then we have for the solutions of the initial value problem (2.3) *the estimate:*

$$x(t) - y(t) = O(\varepsilon) \text{ as } \varepsilon \to 0 \text{ on the timescale } 1/\varepsilon.$$

Remark 2.1 The proof of Theorem 2.1 is rather complicated. Although averaging has been in use for a very long time, the first proof had to wait till 1928 and was given by Fatou [24]. Later proofs were given in [13] and [54]. The complications in the proof are caused by the fact that we have to use an approximating equation which contains terms of the same order in the small parameter ε as the original variational equation. See [58, ch. 2] or [76, ch. 11.3].

For better understanding second order averaging in Chap. 4 we discuss transformations that are used in modern proofs. We start with a near-identity transformation $x \mapsto z$ for Eq. (2.1):

$$x(t) = z(t) + \varepsilon u(t, z(t)) \text{ with } u(t, x) = \int_{t_0}^{t} \left(f(s, x) - f^0(x) \right) ds. \tag{2.4}$$

The variable z is implicitly defined by Eq. (2.4). As we subtracted $f(s, x) - f^0(x)$ the integral will be uniformly bounded (think of the T-periodic $f(t, x)$ as a Fourier series in t where we subtract the 'constant' part $f^0(x)$). Substitution of $x(t)$ into Eq. (2.1) leads after some manipulations to the equation

$$\dot{z} + \varepsilon \frac{\partial}{\partial z} u(t, z)\dot{z} = f^0(z) + O(\varepsilon^2).$$

Details of the calculations can for instance be found in [76]. We have given the expressions to show that $\frac{\partial}{\partial z} u(t, z)$ is expected to play a part in second order approximations (Chap. 4). The near-identity transformation (2.4) is in [58] related to normalisation theory.

Remark 2.2 Condition 4 in Theorem 2.1 uses the notion of 'interior subset'. Such a set is a subset of domain D as used in Theorem 2.1 that consists of the points that are at $O(1)$ distance with respect to ε from the boundary of D. The reason for this condition is that it may happen that the initial conditions of the solution are close to a manifold that separates oscillating and escaping solutions. A small perturbation may push the solution into the domain where solutions have unwanted behaviour.

In addition we have the problem of the interval of time where the solution exists. For linear equations with continuous coefficients the domain of existence extends to all time, $t > t_0$, for nonlinear equations the interval of time remains a point of discussion. We give a simple example.

Example 2.1 Consider the first order equation with initial value $x(0) = 1$:

$$\dot{x} = \varepsilon 2 \cos^2(t) x^2. \tag{2.5}$$

Averaging over t produces the averaged equation

$$\dot{y} = \varepsilon y^2,$$

with solution

$$y(t) = \frac{1}{1 - \varepsilon t}.$$

The function $y(t)$ approximates $x(t)$ with error $O(\varepsilon)$ on intervals of time of size q/ε with for q the condition $0 < q < 1$. The domain D remains of finite size $O(1)$ on these intervals.

Remark 2.3 When introducing and using constants we assume always that they are independent of ε. This also holds for the constant involving the boundedness of the interior subset of domain D. It is important to note that the finite dimension n of the configuration space plays no part in the estimate, the condition is only that the solution starting in x_0 does not leave the subset. *This opens the possibility of averaging very large systems as long as the solutions are bounded.*

Remark 2.4 In the sequel we will often put $t_0 = 0$. Keeping t_0 as a parameter can be useful when connecting different asymptotic approximations in neighbouring domains.

2.2 Quasi-Periodic Averaging

A simple extension of Theorem 2.1 is the case that the vector field $f(t, x)$ in system (2.1) is of the quasi-periodic form:

$$f(t, x) = \sum_{i=1}^{N} f_i(t, x), \tag{2.6}$$

with the vectorfields $f_i(t, x), i = 1, \ldots, N$ periodic with periods $T_i, i = 1, \ldots, N$ independent of each other (incommensurable). We introduce the averaged vectorfield:

$$f^0(x) = \sum_{i=1}^{N} \frac{1}{T_i} \int_{t_0}^{t_0+T_i} f_i(t, x)dt. \tag{2.7}$$

This leads to the theorem:

Theorem 2.2 *Consider the initial value problem (2.1) and the initial value problem with quasi-periodic $f(t, x)$ as in (2.6). We have for the solutions of the initial value problems for x and y the estimate:*

$$x(t) - y(t) = O(\varepsilon) \text{ as } \varepsilon \to 0 \text{ on the timescale } 1/\varepsilon.$$

2.3 Applications

It follows from Theorem (2.1) that the results in Chap. 1 for Eqs. (1.9), (1.12) and Example 1.5 (Mathieu) represent $O(\varepsilon)$ approximations on the timescale $1/\varepsilon$. We will study now a few more problems suitable for periodic averaging, in particular a few iconic examples. A few applications have more than two dimensions. In such a case visualisations are always more difficult, we can use time-series, projections or some geometric features if applicable. Other applications of first order averaging will appear in subsequent chapters.

2.3.1 A Linear Example with Forcing

A well-known problem is to add forcing to the damped harmonic equation. This enables us to show how averaging handles linear resonance.

Example 2.2 The equation is externally excited with $O(\varepsilon h)$ forcing in the form:

$$\ddot{x} + \mu\varepsilon\dot{x} + x = \varepsilon h \sin(\omega t), \tag{2.8}$$

with initial values $x(0), \dot{x}(0)$ given; the constants μ, h, ω are positive. The term $\varepsilon h \sin(\omega t)$ represents a periodic external force on a vibrating mass with size 1. We write this system as

$$\ddot{x} + x = -\varepsilon\mu\dot{x} + \varepsilon h \sin(\omega t).$$

As we have seen in Chap. 1 in the case of non-autonomous perturbations a different form of variational equations than in amplitude-phase is more effective. We transform $x, \dot{x} \rightarrow y_1, y_2$ following Eq. (1.16):

$$x = y_1 \cos t + y_2 \sin t, \quad \dot{x} = -y_1 \sin t + y_2 \cos t. \tag{2.9}$$

We find the variational system:

$$\dot{y}_1 = -\varepsilon \sin t(\mu y_1 \sin t - \mu y_2 \cos t + h \sin(\omega t)),$$

$$\dot{y}_2 = \varepsilon \cos t(\mu y_1 \sin t - \mu y_2 \cos t + h \sin(\omega t)).$$

Averaging the variational system we will, to avoid too much different notation, use again y_1, y_2 to indicate the approximate variables. If $\omega \neq 1$ we find by averaging over t:

$$\dot{y}_1 = -\frac{\varepsilon}{2}\mu y_1, \quad \dot{y}_2 = -\frac{\varepsilon}{2}\mu y_2,$$

showing damping of the oscillations. The forcing is not effective within error $O(\varepsilon)$. Of course we know this from the exact solution. If $\omega \neq 1$ the result to this order of approximation is the damped solution without forcing:

$$x_h(t) = e^{-\frac{\varepsilon}{2}\mu t}(c_1 \cos t + c_2 \sin t),$$

with c_1, c_2 determined by the initial conditions.

The resonant case is $\omega = 1$ which changes the averaged variational system to:

$$\dot{y}_1 = -\frac{\varepsilon}{2}[\mu y_1 + h], \quad \dot{y}_2 = -\frac{\varepsilon}{2}\mu y_2,$$

leading to the $O(\varepsilon)$ approximation valid on the timescale $1/\varepsilon$:

$$x(t) = x_h(t) - \frac{\varepsilon}{2}ht \cos t + O(\varepsilon).$$

$O(h)$ Forcing

Consider now a change of the problem where the forcing is not $O(\varepsilon)$. The equation becomes:

$$\ddot{x} + \mu\varepsilon\dot{x} + x = h \sin(\omega t), \tag{2.10}$$

We know that the solutions for $\varepsilon = 0$ are unbounded if $\omega = 1$ or very large if ω is ε-close to 1. So assume $\omega \neq 1$. A particular solution of the unperturbed equation is with this assumption:

$$\eta(t) = \frac{h}{1 - \omega^2} \sin \omega t.$$

Transforming

$$x = r \cos(t + \phi) + \frac{h}{1 - \omega^2} \sin \omega t, \; \dot{x} = -r \sin(t + \phi) + \frac{\omega h}{1 - \omega^2} \cos \omega t$$

we find with (1.28) the variational equations:

$$\begin{cases} \dot{r} & = \varepsilon \sin(t + \phi) \left(-r \sin(t + \phi) + \frac{\omega h}{1 - \omega^2} \cos \omega t \right), \\ \dot{\phi} & = \varepsilon \cos(t + \phi) \left(-r \sin(t + \phi) + \frac{\omega h}{1 - \omega^2} \cos \omega t \right). \end{cases} \tag{2.11}$$

If $\omega \neq 1$ is rational, the righthand side of the variational system (2.11) is periodic, if ω is irrational the system is quasi-periodic. In both cases averaging produces:

$$\dot{r} = -\frac{\varepsilon}{2} r, \dot{\phi} = 0$$

where we kept r, ϕ for the approximate quantities. The approximate $r(t)$ tends to zero so we conclude

$$x(t) \rightarrow \frac{h}{1 - \omega^2} \sin \omega t, \; \omega \neq 1,$$

a result known for the exact solution.

2.3.2 The Van der Pol-Equation

As a classical example of nonlinear oscillations we present the *Van der Pol-equation*. The equation is basically analysed by first order averaging, multiple timing with t and εt produces the same results in this case.

Example 2.3 The Van der Pol-equation was developed to model electronic oscillations, it was also used to mimic heart-beat oscillations. It is a standard example of so-called self-excited oscillations. The equation is:

$$\ddot{x} + x = \varepsilon \dot{x}(1 - x^2). \tag{2.12}$$

The term $\varepsilon \dot{x}(1 - x^2)$ is called the self-excitation term. We obtain variational equations by using the transformation (1.6) $x, \dot{x} \rightarrow r, \phi$ for a perturbed harmonic equation (1.5). We find the slowly-varying variational system:

$$\begin{cases} \dot{r} & = \varepsilon r \sin^2(t + \phi)(1 - r^2 \cos^2(t + \phi)), \\ \dot{\phi} & = \varepsilon \sin(t + \phi) \cos(t + \phi)(1 - r^2 \cos^2(t + \phi)). \end{cases} \tag{2.13}$$

The righthand side is 2π-periodic in t; averaging over time t produces the system:

$$\dot{r} = \frac{\varepsilon}{2}r(1 - \frac{1}{4}r^2), \; \dot{\phi} = 0.$$

Solving the equations with initial values $r(0) = r_0, \phi(0) = \phi_0$ yields the approximations:

$$r(t) = r_0 \frac{e^{\frac{\varepsilon t}{2}}}{\left(1 + \frac{1}{4}r_o^2(e^{\varepsilon t} - 1)\right)^{\frac{1}{2}}} + O(\varepsilon), \phi(t) = \phi_0 + O(\varepsilon),$$

valid on the timescale $1/\varepsilon$. Taking the limit of $r(t)$ for $t \to \infty$ we find $r(t) \to 2 + O(\varepsilon)$.

As we can see from the averaged equation and also from the approximate $r(t)$, putting $r_0 = 2$ produces as approximation the periodic solution $2\cos(t + \phi_0)$. In Chap. 3 we consider periodic solutions by averaging. See Fig. 2.1 (left).

2.3.3 Averaging Autonomous Equations

Apart from the interesting presence of a periodic solution, the Van der Pol-equation is an example of an autonomous equation, time does not occur explicitly in Eq. (2.12). A more general form of such an autonomous second order ODE is:

$$\ddot{x} + x = \varepsilon f(x, \dot{x}). \tag{2.14}$$

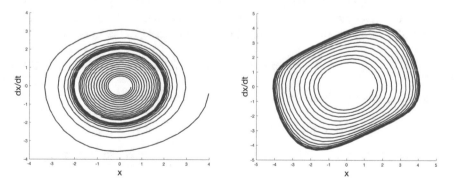

Fig. 2.1 The $x, \dot{x}$ phaseplane of the Van der Pol-equation (left) with $\varepsilon = 0.05, x(0) = 0.1, x(0) = 4, \dot{x}(0) = 0$ (2 orbits). Right the phaseplane of the generalised van der Pol-equation with $\varepsilon = 0.05, a = 2, b = -0.6, x(0) = 1.5, \dot{x}(0) = 0$

Using again amplitude-phase transformation (1.6) produces the variational system:

$$\dot{r} = -\varepsilon \sin(t + \phi) f(r \cos(t + \phi), -r \sin(t + \phi)),$$

$$\dot{\phi} = -\frac{\varepsilon}{r} \cos(t + \phi) f(r \cos(t + \phi), -r \sin(t + \phi)). \tag{2.15}$$

Averaging the system over t keeping r, ϕ constant we note that, because of the autonomous character of the equation, we can as well put $s = t + \phi$ and average over s. We find:

$$-\frac{1}{2\pi} \int_0^{2\pi} \sin(t + \phi) f(r \cos(t + \phi), -r \sin(t + \phi)) dt = F_1(r),$$

$$-\frac{1}{2\pi} \int_0^{2\pi} \cos(t + \phi) f(r \cos(t + \phi), -r \sin(t + \phi)) dt = F_2(r),$$

and so for the averaging approximations:

$$\dot{r} = \varepsilon F_1(r), \quad \dot{\phi} = \varepsilon F_2(r). \tag{2.16}$$

The implication is that averaging of an autonomous second order equation reduces the system to one dimension as solving the equation for r leads to direct integration of the equation for ϕ. We will meet such reductions also for higher dimensional equations.

2.3.4 A Generalised Van der Pol-Equation

In the Van der Pol-equation the self-excitation term $(1 - x^2)\dot{x}$ produces alternating positive and negative damping if $x^2(t)$ becomes alternating larger and smaller than 1. For the value near $r = 2$ this results in a periodic solution. We consider a straightforward generalisation to see what happens. Consider the equation:

$$\ddot{x} + x = \varepsilon \dot{x}(1 - ax^2 - b\dot{x}^2). \tag{2.17}$$

The variational equations obtained by the transformation (1.6) become:

$$\dot{r} = \varepsilon r \sin^2(t + \phi)(1 - ar^2 \cos^2(t + \phi) - br^2 \sin^2(t + \phi)),$$

$$\dot{\phi} = \varepsilon \sin(t + \phi) \cos(t + \phi)(1 - ar^2 \cos^2(t + \phi) - br^2 \sin^2(t + \phi)).$$

The righthand side is 2π-periodic in t; averaging over time t produces the system:

$$\dot{r} = \frac{\varepsilon}{2}r(1 - \frac{a}{4}r^2 - \frac{3}{4}br^2), \ \dot{\phi} = 0.$$

If $a + 3b > 0$ we have an equilibrium at $r = 2/\sqrt{a + 3b}$. In Chap. 3 it will be shown that the equilibrium corresponds with a periodic solution. See Fig. 2.1 (right).

2.3.5 The Mathieu-Equation with Damping

In Example 1.5 we showed that the trivial solution of the Mathieu-equation at prime resonance is unstable. We will consider this problem now when damping is added.

Example 2.4 The damped Mathieu-equation is in this case:

$$\ddot{x} + \varepsilon\mu\dot{x} + (1 + 2\varepsilon \cos 2t)x = 0, \ \mu > 0. \tag{2.18}$$

The parameter product $\varepsilon\mu$ is the scaled damping coefficient. The question is: can the damping neutralise the parametric excitation of the trivial solution?

We use again the transformation (1.16) to co-varying coordinates. We find the variational system:

$$\begin{cases} \dot{y}_1 &= \varepsilon\mu \sin t(-y_1 \sin t + y_2 \cos t) + 2\varepsilon \sin t \cos 2t(y_1 \cos t + y_2 \sin t), \\ \dot{y}_2 &= -\varepsilon\mu \cos t(-y_1 \sin t + y_2 \cos t) - 2\varepsilon \cos t \cos 2t(y_1 \cos t + y_2 \sin t). \end{cases}$$

$$\tag{2.19}$$

Averaging the 2π-periodic system produces:

$$\dot{y}_1 = -\varepsilon\frac{\mu}{2}y_1 - \frac{1}{2}\varepsilon y_2, \ \dot{y}_2 = -\varepsilon\frac{\mu}{2}y_2 - \frac{1}{2}\varepsilon y_1. \tag{2.20}$$

The eigenvalues near $(0, 0)$ are $\frac{\varepsilon}{2}(-\mu \pm 1)$ so that the trivial solution is stable if $\mu > 1$.

2.3.6 The Forced Duffing-Equation Without Damping

The harmonic equation has a family of 2π-periodic solutions, the equation is called isochronous. Forcing the harmonic equation with a 2π-periodic function leads to unbounded solutions (linear resonance), forcing with a different period leads to oscillation with bounded amplitude. What happens if we consider this equation with a cubic term added, the Duffing-equation.

So consider the forced Duffing-quation:

$$\ddot{x} + x + \varepsilon\gamma x^3 = \varepsilon h \cos t, \qquad (2.21)$$

with $\gamma, h \neq 0$ constants. Transforming with (1.6) we find in amplitude-phase variables r, ϕ the variational system and from this the averaged system:

$$\dot{r} = -\frac{1}{2}\varepsilon h \sin\phi,$$

$$\dot{\phi} = -\frac{1}{2}\varepsilon\left(-\frac{3}{4}\gamma r^2 + h\frac{\cos\phi}{r}\right).$$

Stationary solutions (r_0, ϕ_0), $r_0 > 0$ satisfy the transcendental equations

$$h \sin\phi_0 = 0, \quad -\frac{3}{4}\gamma r_0^2 + h\frac{\cos\phi_0}{r_0} = 0.$$

If $\sin\phi_0 = 0$, we have $\cos\phi_0 = \pm 1$. This produces for the stationary solution the condition:

$$r_0^3 = \mp\frac{4h}{3\gamma}. \qquad (2.22)$$

We are looking for a positive solution of r_0. The corresponding approximation for $x(t)$ is valid with error $O(\varepsilon)$ on the timescale $1/\varepsilon$.

We will return later to the question whether the approximation describes a periodic solution. In general we will discuss later on the interesting question whether a perturbation result is qualitatively persistent under higher order perturbations.

2.3.7 Nonlinear Damping

If the damping is progressive, for instance in the equation:

$$\ddot{x} + \varepsilon(\dot{x} + \dot{x}^3) + x = 0,$$

we can see that damping to zero equilibrium is stronger than in the linear case. This can be shown by averaging but also by differentiation of the so-called action $I = \frac{1}{2}(\dot{x}^2 + x^2)$. We find

$$\frac{dI}{dt} = \dot{x}\ddot{x} + x\dot{x} = -\varepsilon(\dot{x}^2 + \dot{x}^4).$$

The action decreases stronger than for the linear case.

What happens if the damping weakens?

Example 2.5 Consider the equation:

$$\ddot{x} + \varepsilon(\dot{x} - \dot{x}^3) + x = 0. \tag{2.23}$$

Transformation (1.6) produces the variational equations:

$$\begin{cases} \dot{r} & = -\varepsilon \sin(t + \phi)(r \sin(t + \phi) - r^3 \sin^3(t + \phi)), \\ \dot{\phi} & = -\frac{\varepsilon}{r} \cos(t + \phi)(r \sin(t + \phi) - r^3 \sin^3(t + \phi)). \end{cases} \tag{2.24}$$

Averaging over t produces;

$$\dot{r} = -\frac{\varepsilon}{2} r (1 - \frac{3}{4} r^2), \ \dot{\phi} = 0.$$

We find 2 equilibria, $r = 0$ and $r = 2/\sqrt{3}$. This is qualitatively a new phenomenon but by linearisation near the equilibria we find that the trivial solution is stable, the 2nd equilibrium is unstable. For the stability by linearisation in more general theory one can consult [76].

2.3.8   A Hamiltonian System with Cubic Term

The results in this subsection are based on [74]. Consider a two degrees-of-freedom (dof) Hamiltonian system generated by the Hamiltonian:

$$H(p, q) = \frac{1}{2}(p_1^2 + \omega^2 q_1^2) + \frac{1}{2}(p_2^2 + q_2^2) - \varepsilon q_1 q_2^2. \tag{2.25}$$

The variables p_1, p_2 are called the momenta (here identified with velocities), the variables q_1, q_2 are called positions. The quadratic part of the Hamiltonian is often abbreviated as $H_2(p, q)$, ω and 1 are frequencies in the linear part of the equations of motion, in this example we choose $\omega = 2$. The equations of motion are derived from:

$$\dot{q}_i = \frac{\partial H}{\partial p_i}, \ \dot{p}_i = -\frac{\partial H}{\partial q_i}, i = 1, 2.$$

The equations are for this Hamiltonian:

$$\begin{cases} \ddot{q}_1 + 4q_1 & = \varepsilon q_2^2, \\ \ddot{q}_2 + q_2 & = 2\varepsilon q_1 q_2. \end{cases} \tag{2.26}$$

The equations of motion have been used as a cartoon problem for the motion of stars near the galactic plane in a rotating disk galaxy. The energy of a particle moving according to system (2.26) is given by $H(p, q)$, the energy manifolds in 4-dimensional phase-space are determined by $H(p, q) = $ constant. For a general introduction to integrals of ODEs and the corresponding integral manifolds see [76]. These manifolds (surfaces in 4-dimensional phasespace) are close to ellipsoïds, they are bounded (compact) in a sphere around the origin of radius size $1/\varepsilon$. Within this large sphere, but in a neighbourhood of the origin, according to KAM-theory, see [18], there exist an infinite family of tori that surround an infinite number of periodic solutions. For larger values of the energy but still on bounded energy manifolds chaos will emerge.

The problem for system (2.26) looks simple but the dynamics is complicated. We will see what first order averaging has to say about this problem. Slowly varying equations arise from transformation (1.6) $(q_i, \dot{q}_i \mapsto r_i, \psi_i, i = 1, 2)$:

$$
\begin{cases}
\dot{r}_1 &= -\frac{1}{2}\varepsilon \sin(2t + \psi_1) r_2^2 \cos^2(t + \psi_2), \\
\dot{\psi}_1 &= -\varepsilon \frac{\cos(2t + \psi_1)}{2r_1} r_2^2 \cos^2(t + \psi_2), \\
\dot{r}_2 &= -2\varepsilon \sin(t + \psi_2) r_1 r_2 \cos(t + \psi_2) \cos(2t + \psi_1), \\
\dot{\psi}_2 &= -2\varepsilon r_1 \cos(2t + \psi_1) \cos^2(t + \psi_2).
\end{cases}
\tag{2.27}
$$

Note that we have to exclude regions where the amplitudes are small or zero. The 2π-periodicity of the righthand sides suggests averaging over t. To avoid too many indices we use as before the original variables for the approximate quantities. Averaging yields:

$$
\begin{cases}
\dot{r}_1 &= -\frac{\varepsilon}{8} r_2^2 \sin \chi, \quad \dot{\psi}_1 = -\frac{\varepsilon}{8} \frac{r_2^2}{r_1} \cos \chi, \\
\dot{r}_2 &= -\frac{\varepsilon}{2} r_1 r_2 \sin \chi, \quad \dot{\psi}_2 = -\frac{\varepsilon}{2} r_1 \cos \chi.
\end{cases}
\tag{2.28}
$$

We introduced the combination angle $\chi = \psi_1 - 2\psi_2$. Our aim is to find periodic solutions and invariant manifolds embedded on the energy manifolds characterising the phaseflow.

Periodic Solutions

Note that putting $q_2 = \dot{q}_2 = 0$ in system (2.26) (a q_1 normal mode) produces a family of periodic solutions, parameterised by the energy. This q_1-normal mode family is not described by the averaged equations in amplitude-phase form. An exact q_2-normal mode family does not exist but is there such an ε-close family? From system (2.28) we find that r_1 is forced by $r_2 \neq 0$ so this seems not the case. The amplitude-phase transformation is of course not suitable for normal mode analysis but repeating the calculations with transformation (1.16) we find again that there is not a q_2-normal mode family. This is surprising as one expects often continuation from the normal modes of the linearised system to the nonlinear case.

Periodic solutions in *general position* (not ε-close to coordinate planes) arise from system (2.28) if $\sin \chi = 0$ or $\chi = 0, \pi$. These solutions with constant amplitude exist if $\dot{\chi} = 0$ so that:

$$\dot{\chi} = \dot{\psi}_1 - 2\dot{\psi}_2 = -\varepsilon \left(\frac{r_2^2}{8r_1} - r_1 \right) \cos \chi = 0,$$

leading to the requirement

$$r_2^2 = 8r_1^2.$$

This yields 2 families of periodic solutions; we can make them more explicit by considering them on invariant manifolds.

Invariant Manifolds

Multiplying in system (2.28) the equation for r_1 with r_1, the equation for r_2 with r_2, adding the equations and integrating we find the approximate energy integral:

$$4r_1^2 + r_2^2 = 2E_0, \tag{2.29}$$

with E_0 a positive constant, the initial energy. This is in fact an $O(\varepsilon)$ approximation of the energy integral $H = $ constant. In the case of bounded solutions of the Hamiltonian system we do not need averaging to obtain result (2.29) from Hamiltonian (2.25). Combining the condition for general position periodic solutions with integral (2.29) we find:

$$r_1 = \sqrt{\frac{E_0}{6}}, \ r_2 = \sqrt{\frac{4E_0}{3}}.$$

The 2 families are parameterised by the energy.

Fig. 2.2 Poincaré map of the 2 : 1 resonance Hamiltonian system described by (2.28); the energy is fixed and the orbits move on a fixed energy manifold puncturing the plane $\dot{q}_2 = 0$. The fixed point in the centre corresponds with the unstable q_1 normal mode. The closed orbits consist of points where the orbits are passing the transversal plane, the orbits are fiilling up tori

Remarkably enough we find a second integral from system (2.28):

$$\frac{1}{2}r_1 r_2^2 \cos \chi = I, \tag{2.30}$$

with I a constant. The 2 integrals and the 3 families of periodic solutions are useful to obtain a picture of the geometry of phase-space to $O(\varepsilon)$. The analysis for the following summary is based on the study of the averaged system (2.28); we summarise:

The general position orbits are Lyapunov stable on each energy manifold, the q_1-normal mode family is unstable. For definitions of various types of stability see [76]. One can visualise the flow on a fixed energy manifold but this flow is still 3-dimensional. It helps to consider a transversal of this 3-dimensional flow for a fixed value of one of the variables; this corresponds with a plane that is punctured by the oscillating orbits on the energy manifold and is called a Poincaré map. The q_1 normal mode produces 1 point, a saddle, see Fig. 2.2. Because of the frequency 2 of the q_1 mode the periodic orbits produce 2 fixed points each in the transversal map. The closed curves around the fixed points consist actually of points from orbits on the tori surrounding the stable periodic solutions that puncture the transversal plane. See for details [74].

2.3.9 The Spring-Pendulum

Consider a hanging spring attached at the top and with a weight attached to the lower end, see Fig. 2.3 . The spring can oscillate without bending and it can swing in a vertical plane as a pendulum making an angle ϕ with the vertical; without friction it can be described by a Hamiltonian system with 2 dof. A model extension would be the *spherical* spring-pendulum requiring 3 dof. We will study the stability

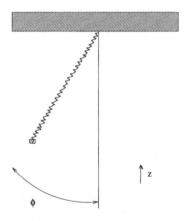

Fig. 2.3 The spring-pendulum described by system (2.32)

of the normal mode vertical oscillations in the 2 dof case. In 1933 Gorelik and Vitt [29] showed experimentally that the vertical normal mode can be unstable resulting in strong exchanges of energy between the 2 dof. In [70] this phenomenon was explained by first order averaging. In section 10.5 of [28] the role of symmetries in the dynamics of the spring-pendulum is analysed.

A local nonlinear analysis involves expanding the Hamiltonian to a certain order and rescaling the variables by ε to study the dynamics near equilibrium when the spring is hanging down at rest. For the expanded Hamiltonian $H = H_2 + \varepsilon H_3$ we have from [70] for small deflections to degree 3:

$$\begin{cases} H_2 &= \frac{1}{2}\omega_z(p_z^2 + z^2) + \frac{1}{2}\omega_\phi(p_\phi^2 + \phi^2), \\ H_3 &= \frac{\omega_\phi}{\sqrt{ml_0^2\omega_z}}\left(\frac{1}{2}z\phi^2 - zp_\phi^2\right). \end{cases} \tag{2.31}$$

Note that we have mirror (discrete) symmetry in ϕ as expected. To avoid too many parameters we put $ml_0^2 = 1$. Consider the resonance ratio by choosing $\omega_z = 2$, $\omega_\phi = 1$. The equations of motion become to $O(\varepsilon)$:

$$\begin{cases} \dot{z} &= 2p_z, \\ \dot{p}_z &= -2z - \frac{\varepsilon}{\sqrt{2}}(\frac{1}{2}\phi^2 - p_\phi^2), \\ \dot{\phi} &= p_\phi(1 - \varepsilon\sqrt{2}z), \\ \dot{p}_\phi &= -\phi - \frac{\varepsilon}{\sqrt{2}}z\phi. \end{cases} \tag{2.32}$$

The normal mode of vertical motion of the spring ($\dot{\phi} = \dot{p}_\phi = 0$) is a solution of system (2.32), in fact a harmonic function. We will study the stability of this normal mode.

The equation for $\dot{\phi}$ enables us to eliminate p_ϕ, the equations of motion can to $O(\varepsilon)$ be written as:

$$\ddot{z} + 4z = \varepsilon\frac{\sqrt{2}}{2}(\dot{\phi}^2 - \frac{1}{2}\phi^2), \quad \ddot{\phi} + \phi = \varepsilon\sqrt{2}(\frac{1}{2}z\phi - \dot{z}\dot{\phi}). \tag{2.33}$$

We can apply transformation (1.6): $z = r_1\cos(2t + \psi_1)$, $\phi = r_2\cos(t + \psi_2)$ etc. although considering ϕ small to study the stability of the vertical normal mode makes this questionable. However, because of the nonlinear terms in the equations for the ϕ-mode the polar coordinate singularity is in this case neutralised. Introducing the combination angle $\chi = \psi_1 - 2\psi_2$ we have by first order averaging for the variational equations:

$$\begin{cases} \dot{r}_1 &= \varepsilon\frac{3\sqrt{2}}{32}r_2^2\sin\chi, \quad \dot{\psi}_1 = \varepsilon\frac{3\sqrt{2}}{32}\frac{r_2^2}{r_1}\cos\chi, \\ \dot{r}_2 &= -\varepsilon\frac{3\sqrt{2}}{8}r_1r_2\sin\chi, \quad \dot{\psi}_2 = \varepsilon\frac{3\sqrt{2}}{8}r_1\cos\chi. \end{cases} \tag{2.34}$$

For the combination angle χ we have the equation:

$$\dot{\chi} = \varepsilon \frac{3}{4}\sqrt{2}(\frac{r_2^2}{8r_1} - r_1) \cos \chi.$$ (2.35)

System (2.34) has the first integral:

$$4r_1^2 + r_2^2 = 2E_0,$$ (2.36)

with E_0 a positive constant. Choosing $r_2^2 - 8r_1^2 = 0$ we have $\dot{\chi} = 0$; this relation holds for $t \geq 0$ if $r_1(t), r_2(t)$ are constant, so choose $\chi = 0, \pi$. This results in 2 families of periodic solutions. Using integral (2.36) we find with $\chi = 0, \pi$:

$$r_1^2 = \frac{1}{6}E_0, \ r_2^2 = \frac{4}{3}E_0.$$ (2.37)

This makes with the vertical normal mode 3 families of periodic solutions. In Fig. 2.2 we used a Poincaré map for illustration of the Hamiltonian flow on the energy manifold. As we can reduce the analysis in the case of the spring-pendulum, we can also use a reduction to the r_1, χ-plane; see Fig. 2.4.

For $\varepsilon = 0$ we have the pendulum normal mode in system (2.32) but this mode can not be continued.

In studies of mechanical oscillations with 2 dof one often finds the $1 : 2$ resonance, both in conservative and in dissipative systems. The main reason is that a first order approximation in ε is usual sufficient in these problems. Other resonance ratios will be considered in Chap. 8, they require more advanced perturbation methods.

Fig. 2.4 The r_1, χ phase-plane of the spring-pendulum described by system (2.34); we have chosen $E_0 = 1$ so we have from (2.36) that $0 \leq r_1 \leq 0.707$. The solutions corresponding with $\chi = 0, \pi$ are stable, the vertical normal mode corresponding with $r_1 = 0.707, r_2(t) = 0$ is unstable

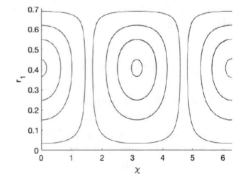

Chapter 3
Periodic Solutions

A mathematical theory of approximation should, because of its guaranteed precision, also produce *qualitative* statements. In other words: if we find a phenomenon by approximation does it persist for the original problem with the original equations? We will discuss a few basic cases in various chapters. The next theorem on the existence of periodic solutions by averaging is sometimes called "Bogoliubov's second theorem". We will see in Chap. 9 that the extension to the presence of tori in dynamical systems has a relation with the Bogoliubov theory of periodic solutions.

In addition we will consider the Poincaré-Lindstedt method that we met already in the introductory Chap. 1. This useful result is, like the second Bogoliubov theorem, based on the implicit function theorem but of a different type as it leads to *convergent* approximations. As we shall see the averaging aspects are hidden in a periodicity condition.

3.1 Periodic Solutions

Equilibria of averaged variational systems.

Theorem 3.1 *Consider the variational n-dimensional system (2.1)* $\dot{x} = \varepsilon f(t, x)$ *with* $f(t, x)$ *T-periodic in t and the averaged system* $\dot{y} = \varepsilon f^0(y)$. *Suppose that the conditions of theorem 2.1 are satisfied. Assume in addition that* y_0 *is a stationary solution (critical point) of the averaged equation,* $f^0(y_0) = 0$. *If*

1. $f(t, x)$ *is a smooth vector field;*
2. *for the determinant of the Jacobian J in* y_0 *we have*

$$\left| \frac{\partial f^0(y)}{\partial y}\big|_{y=y_0} \right| \neq 0,$$

© The Author(s), under exclusive license to Springer Nature Switzerland AG 2023
F. Verhulst, *A Toolbox of Averaging Theorems*, Surveys and Tutorials in the Applied
Mathematical Sciences 12, https://doi.org/10.1007/978-3-031-34515-9_3

then a T-periodic solution of the equation $\dot{x} = \varepsilon f(t, x)$ *exists in an ε-neighbourhood of $x = y_0$.*

Remark 3.1 A *proof* can be found in [13] and in [76, ch. 11.8] based on the implicit function theorem. The use of this theorem means that the periodic solutions obtained in this way have a period close to T and amplitudes close to the equilibrium value. Note that apart from the periodic solutions obtained from the equilibria the variational system may obtain other periodic solutions.

We can even establish the stability of the periodic solution, as it matches exactly the stability of the stationary solution of the averaged equation if the real parts of the eigenvalues are nonzero. This reduces the stability problem of the periodic solution to determining the eigenvalues of a matrix.

Later we will meet the notion of hyperbolic and hyperbolicity. A $n \times n$ matrix is called structurally stable if the n eigenvalues have all of them a nonzero real part. We call this also *hyperbolicity* of the critical point.

Remark 3.2 Following the error estimate of theorem (2.1) the error of the approximation is $O(\varepsilon)$ on the timescale $1/\varepsilon$. However, as we have existence of the periodic solution we have for the amplitude validity of the error estimate for all time.

Remark 3.3 We have seen before that when starting with an *autonomous* equation or system and averaging the corresponding variational equations one eliminates one angle or, more generally, one reduces the system with one dimension. The implication is that a stationary solution obtained from an autonomous equation always has one eigenvalue zero and so the determinant of J, $det(J) = 0$. This is for instance the case for the Van der Pol-equation. In accordance with this observation, we have an important modification of the existence requirement for periodic solutions in the case of n-dimensional autonomous equations: the Jacobian in the equlibrium y_0 in Theorem 3.1 must have rank $n - 1$.

The eigenvalue zero is a general feature of solutions of autonomous equations, see [76].

3.2   Applications

3.2.1   A First Order Equation

Consider the variational equation:

$$\dot{x} = \varepsilon(x - 2x^2 \sin^2 t). \tag{3.1}$$

Averaging over t produces:

$$\dot{y} = \varepsilon(y - y^2).$$

The 2 equilibria of the averaged equation are $y_0 = 0, 1$. We have from Theorem 3.1:

$$\frac{\partial f^0(y)}{\partial y}|_{y=y_0} = 1 - 2y_0.$$

For both equilibria the Jacobian does not vanish, we conclude that both equilibria have an ε-close 2π-periodic solution. It is easy to see that for $y_0 = 0$ the periodic solution is unstable, near $y_0 = 1$ stable.

In this case we can compute the stable periodic solution $x_s(t)$ by variation of constants exactly:

$$x_s(t) = \frac{4 + \varepsilon^2}{4 + \varepsilon^2 - 2\varepsilon \sin 2t - \varepsilon^2 \cos 2t}. \tag{3.2}$$

3.2.2 A Generalised Van der Pol-Equation

Consider an example of the equations formulated in Chap. 1:

$$\ddot{x} + x = \varepsilon\dot{x}(1 - 3x^2 - 5\dot{x}^2). \tag{3.3}$$

Introducing amplitude-phase coordinates by transformation (1.6) and averaging produces:

$$\dot{r} = \frac{\varepsilon}{2}r(1 - \frac{9}{2}r^2), \quad \dot{\psi} = 0. \tag{3.4}$$

We can conclude that $r = \sqrt{2}/3$ is a stable critical point of the averaged equation for r. Applying Theorem 3.1 and Remark 3.3 we conclude that Eq. (3.3) contains an asymptotically stable periodic solution with approximation

$$x(t) = \frac{\sqrt{2}}{3}\cos(t + \psi(0)) + O(\varepsilon).$$

3.2.3 The Duffing-Equation with Small Forcing

Consider again the Duffing-equation with small forcing and damping, ω near 1:

$$\ddot{x} + \varepsilon\mu\dot{x} + \varepsilon\gamma x^3 + x = \varepsilon h \cos\omega t,$$

parameters $\mu \geq 0, h \neq 0, \omega^{-2} = 1 - \beta\varepsilon$. This is a typical mechanical example representing nonlinear oscillations with damping and forcing close to resonance.

Transforming $\tau = \omega t$ we have in amplitude-phase variables the averaged system

$$r' = -\frac{1}{2}\varepsilon(\mu r + h \sin \psi),$$

$$\psi' = -\frac{1}{2}\varepsilon\left(\beta - \frac{3}{4}\gamma r^2 + h\frac{\cos \psi}{r}\right).$$

Stationary solutions (equilibria) $r_0 > 0$, ψ_0 satisfy the transcendental equations

$$\mu r_0 + h \sin \psi_0 = 0, \; \beta - \frac{3}{4}\gamma r_0^2 + h\frac{\cos \psi_0}{r_0} = 0.$$

For the derivative of the right-hand side of the averaged system, leaving out the scalefactor $-\frac{1}{2}\varepsilon$, we find the Jacobian at r_0, ψ_0 ($r_0 > 0$) :

$$J = \begin{pmatrix} \mu & h \cos \psi_0 \\ -\frac{3}{2}\gamma r_0 - h\frac{\cos \psi_0}{r_0^2} & -h\frac{\sin \psi_0}{r_0} \end{pmatrix}.$$

Leaving out the scalefactor $\frac{1}{4}\varepsilon^2$, the determinant of the Jacobian J becomes at equilibrium:

$$det(J) = \left(-\mu \sin \psi_0 + \frac{3}{2}\gamma r_0^2 \cos \psi_0 + h\frac{\cos^2 \psi_0}{r_0}\right)\frac{h}{r_0}.$$

For periodic solutions to exist, we have from Theorem 3.1 the requirement $|J| \neq 0$ if the stationary solutions are substituted.

To be more explicit, we will look at the case of exact resonance $\beta = 0$. For the stationary solutions (equilibria), we have

$$h \sin \psi_0 = -\mu r_0, \; h \cos \psi_0 = \frac{3}{4}\gamma r_0^3.$$

Using that $\cos^2 \psi_0 + \sin^2 \psi_0 = 1$, we find

$$\frac{9}{16}\gamma^2 r_0^6 + \mu^2 r_0^2 - h^2 = 0,$$

which always admits at least one positive solution for r_0. For the Jacobian, we have, eliminating ψ_0,

$$det(J) = r_0\left(\mu + \frac{27}{16}\gamma^2 r_0^4\right).$$

If μ and γ do not vanish simultaneously, we have $det(J) > 0$, so a stationary solution $r_0 > 0$, ψ_0 corresponds with a periodic solution of the original forced

Duffing-equation. The stability of the periodic solution follows from the eigenvalues of the equiibrium and depends on the values of the parameters.

3.2.4 The Nonlinear Mathieu-Equation

Consider the nonlinear Mathieu-equation:

$$\ddot{x} + (1 + \varepsilon \cos 2t)x + \varepsilon a x^3 = 0.$$

From amplitude-phase transformation (1.6) we find for the variational system:

$$\dot{r} = \varepsilon \sin(t + \psi)[\cos 2t \, r \cos(t + \psi) + a r^3 \cos^3(t + \psi)],$$
$$\dot{\psi} = \frac{\varepsilon}{r} \cos(t + \psi)[\cos 2t \, r \cos(t + \psi) + a r^3 \cos^3(t + \psi)].$$

Averaging produces, excluding a neighbourhood of the origin:

$$\dot{r} = \frac{1}{4}\varepsilon r \sin 2\psi, \quad \dot{\psi} = \frac{1}{4}\varepsilon(\cos 2\psi + \frac{3}{2}ar^2). \tag{3.5}$$

This system is based on polar coordinates and not convenient to analyse near $r = 0$, but for positive radius we find stationary r_0, ψ_0 if $\psi_0 = 0, \pi/2, \pi, 3\pi/2$. Two periodic solutions are found for $a > 0$ if $\psi_0 = \pi/2, 3\pi/2$, for $a < 0$ if $\psi_0 = 0, \pi$.

The Jacobian of the righthand sides of system (3.5) is, omtting the scale factor $\frac{1}{4}\varepsilon$:

$$J = \begin{pmatrix} \sin 2\psi_0 & 2r_0 \cos 2\psi_0 \\ 3ar_0 & -2\sin 2\psi_0 \end{pmatrix}.$$

For the determinant we have:

$$det(J) = -2\sin^2 2\psi_0 - 6ar_0^2 \cos 2\psi_0.$$

In both cases of the sign of a we have $det(J) \neq 0$.

If $a > 0$ we find 2 periodic solutions near $r_0 = \sqrt{2a/3}$, if $a < 0$ 2 periodic solutions near $r_0 = \sqrt{-2a/3}$.

To study the behaviour near the trivial solution we use for the variational equations transformation (1.16). We find:

$$\dot{y}_1 = \varepsilon \sin t(\cos 2t(y_1 \cos t + y_2 \sin t) + a(y_1 \cos t + y_2 \sin t)^3),$$
$$\dot{y}_2 = \varepsilon \cos([\cos 2t(y_1 \cos t + y_2 \sin t) + a(y_1 \cos t + y_2 \sin t)^3).$$

Averaging produces

$$\dot{y}_1 = \frac{1}{4}\varepsilon(-y_2 + \frac{3a}{2}y_1^2 y_2 + \frac{3a}{2}y_2^3), \ \dot{y}_2 = \frac{1}{4}\varepsilon(y_1 + \frac{3a}{2}y_1^3 + \frac{3a}{2}y_1 y_2^2).$$

Linearisation produces near $y_1 = y_2 = 0$ a saddle with eigenvalues $\pm\frac{1}{4}\varepsilon$ so the trivial solutions of both the linear and nonlinear Mathieu-equation are unstable.

3.2.5 The Mathieu-Equation with Damping

Consider the Mathieu-equation with damping:

$$\ddot{x} + \varepsilon\kappa\dot{x} + (\omega^2 + \varepsilon \cos 2t)x = 0, \tag{3.6}$$

with $\kappa > 0$ the damping coefficient. The equation without damping was studied in Example 1.5. The behaviour of the solutions depends strongly on the parameters; as an example we choose the frequency ω near 1. This is expressed by putting $\omega^2 = 1 + \varepsilon\beta$ with β a constant independent of ε. The equation is transformed to amplitude-phase variables by transformation Eq. (1.6) and becomes:

$$\dot{r} = \varepsilon(-\kappa r \sin^2(t + \psi) + \beta r \sin(t + \psi)\cos(t + \psi)$$
$$+r \sin(t + \psi)\cos 2t \cos(t + \psi)),$$
$$\dot{\psi} = \varepsilon(-\kappa \sin(t + \psi)\cos(t + \psi) + \beta \cos^2(t + \psi) + \cos^2(t + \psi)\cos 2t).$$

Averaging over time t produces:

$$\dot{r} = -\frac{\varepsilon}{2}r(\kappa - \frac{1}{2}\sin 2\psi), \ \dot{\psi} = \frac{\varepsilon}{2}(\beta + \frac{1}{2}\cos 2\psi). \tag{3.7}$$

In the ω, ε-parameterspace, the stable and unstable solutions are located in different domains, see Fig. 3.1. The domains are separated in parameterspace by curves and as these curves separate stability and instability we infer that the separating curves contain the parameters of the periodic solutions.

To find the separating curves we will look for periodic solutions using the equilibria of averaged system (3.7). We assume that such simple and prominent types of periodic solutions are characterised by constant amplitude r and constant phase ψ. The equilibria follow from the equations:

$$2\kappa = \sin 2\psi, \ 2\beta = -\cos 2\psi.$$

The results from Eq. (3.7) are:

Fig. 3.1 The (ω^2, ε) parameter space of the damped Mathieu-equation for ε small (the scaling of the axes is different). The approximate solutions are described by system (3.7). The instability domain emanating from the horizontal axis in the case without damping is called a Floquet tongue. With damping the tongue is lifted from the horizontal axis

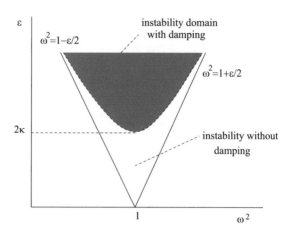

Without damping $\kappa = 0$, choose $\beta = \pm 0.5$. We find $\sin 2\psi = 0$ or $\cos 2\psi = \pm 1$ producing 2 straight lines in parameter-space.:

$$\omega^2 = 1 \pm \frac{1}{2}\varepsilon.$$

With damping we have for stationary r_0: $\sin 2\psi_0 = 2\kappa$; eliminating the phase ψ_0 we find:

$$\kappa^2 + \beta^2 = \frac{1}{4},$$

and by eliminating β:

$$\omega^2 = 1 \pm \frac{\varepsilon}{2}\sqrt{1 - 4\kappa^2}, \quad \kappa^2 \le \frac{1}{4}. \tag{3.8}$$

In the case of damping the curve separating stability and instability in parameterspace is lifted off the horizontal frequency-axis. See Fig. 3.1.

3.2.6 Models for Autoparametric Energy Absorption

In dynamical systems, in general in mathematics, one is interested in vibrational phenomena but in actual engineering problems, we may wish to diminish or even extinguish the vibration amplitudes of a system (called the Primary System). Think of two examples of Primary Systems, a sea-going vessel that is excited by the waves or machinery on a foundation that is slightly excited by neighbouring machinery. A straightforward technique is using energy absorption by using dampers but they have to replaced often and sometimes 'quenching of undesirable vibrations' can

Fig. 3.2 Two coupled
oscillators with vertical
oscillations as primary system
and parametric excitation of
the coupled pendulum
(secondary system)

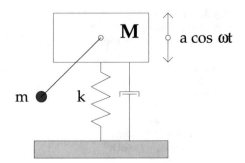

be achieved by coupling the Primary system to a Secondary System that absorbs the energy. A systematic description of such quenching models using the theory of autoparametric systems, can be found in [63].

As the calculations are long we present in the following two applications only a sketch of the results from [63], this will give an impression of these interesting engineering problems.

Example 3.1 (Energy Absorption by an Attached Pendulum) A typical example of autoparametric excitation is shown in Fig. 3.2. The Primary System that we would like to be stationary is the mass M that rests on an elastic foundation and is forced vertically by oscillations with amplitude a and frequency ω. We attach to the mass M a pendulum with mass m as a Secondary System. The vertical (normal mode) vibrations of the Primary System with mass M can (hopefully) act as parametric excitation of the (pendulum) Secondary System which will no longer remain at rest. Such parametric excitation will take place if the parameters of the vertical normal mode of the Primary System are in the instability (parameter) intervals of this normal mode solution in the full, coupled system.

The deflection of mass M in the vertical direction will be indicated by y, for $y = 0$ the mass will be at rest. The pendulum deflections are indicated by the angle ϕ that the pendulum makes with the vertical axis. Following [63] with some rescaling we formulate the equations of motion for small deflections and assumptions on the frequencies (with small detunings σ_1, σ_2) of the vertical oscillations and the pendulum motion:

$$\begin{cases} \ddot{y} + y & = -\varepsilon(\kappa_1 \dot{y} + \sigma_1 y - \tfrac{1}{4}\phi^2 + \mu\dot{\phi}^2 + a\cos t), \\ \ddot{\phi} + \tfrac{1}{4}\phi & = -\varepsilon(\kappa_2 \dot{\phi} + \tfrac{1}{2}\sigma_2\phi - y\phi). \end{cases} \tag{3.9}$$

We have the positive damping coefficients $\kappa_{1,2}$, the detunings $\sigma_{1,2}$ of the basic frequencies $1, 1/2$ and forcing amplitude $a > 0$, also parameter $\mu > 0$. This type of spring-pendulum is a classical problem that was studied by many authors, see for references [63]. We have the normal mode solution $\phi = \dot{\phi} = 0$ corresponding with forced linear y-oscillations of the mass M. We can solve the equation to find a periodic normal mode solution:

$$y_n(t) = -a \frac{\sigma_1}{\kappa_1^2 + \sigma_1^2} \cos t - a \frac{\kappa_1}{\kappa_1^2 + \sigma_1^2} \sin t, \tag{3.10}$$

We introduce slowly-varying variational variables by transformation (1.6) where r_1 is associated with the deflection y, r_2 with ϕ. First order averaging produces for general position orbits:

$$\begin{cases} \dot{r}_1 = \frac{1}{2}\varepsilon(-\kappa_1 r_1 - \frac{1}{8}(1+\mu)r_2^2 \sin\chi + a\sin\psi_1), \\ \dot{\psi}_1 = \frac{1}{2r_1}\varepsilon(\sigma_1 r_1 - \frac{1}{8}(1+\mu)r_2^2 \cos\chi + a\cos\psi_1), \\ \dot{r}_2 = \frac{1}{2}r_2\varepsilon(-\kappa_2 + r_1 \sin\chi), \\ \dot{\psi}_2 = \frac{1}{2}\varepsilon(\sigma_2 - r_1 \cos\chi), \end{cases} \tag{3.11}$$

with phase-difference $\chi = \psi_1 - 2\psi_2$. Putting $\phi = \dot{\phi} = 0$ in system (3.9) we obtain after averaging the first 2 equations of system (3.11) with $r_2 = 0$. Consider now system (3.11) with $r_2 > 0$. Looking for critical points we find from the last 2 equations:

$$\sin\chi = \frac{\kappa_2}{r_1}, \cos\chi = \frac{\sigma_2}{r_1}, r_1 = \sqrt{\kappa_2^2 + \sigma_2^2}. \tag{3.12}$$

From the first 2 equations we find after some manipulations a quadratic equation for r_2^2:

$$(\kappa_1 r_1^2 + \frac{\kappa_2}{8}(1+\mu)r_2^2)^2 + (\sigma_1 r_1^2 - \frac{\sigma_2}{8}(1+\mu)r_2^2)^2 - a^2 r_1^2 = 0.$$

or

$$\frac{1}{64}(1+\mu)^2 r_2^4 + \frac{1}{4}(\kappa_1\kappa_2 - \sigma_1\sigma_2)(1+\mu)r_2^2 + (\kappa_1^2 + \sigma_1^2)(\kappa_2^2 + \sigma_2^2) - a^2 = 0.$$

One can easily deduce from this equation conditions on the parameters to have no, one or two positive solutions for r_2. For instance if $(\kappa_1^2 + \sigma_1^2)(\kappa_2^2 + \sigma_2^2) - a^2 > 0$ and $\kappa_1\kappa_2 - \sigma_1\sigma_2 > 0$, there are no positive solutions. For the reader we leave these details and the stability study to section 4.2 of [63].

As an instructive but even more complicated example we consider briefly the following autoparametric system studied in [23]:

Example 3.2 (Autoparametric Excitation with Complex Dynamics)

$$x'' + x + \varepsilon(k_1 x' + \sigma_1 x + a \cos 2\tau x + \frac{4}{3}x^3 + c_1 y^2 x) = 0$$

$$y'' + y + \varepsilon(k_2 y' + \sigma_2 y + c_2 x^2 y + \frac{4}{3}y^3) = 0$$

(3.13)

where σ_1 and σ_2 are the detunings from the 1 : 1-resonance of the 2 oscillators. In this system, $y(t) = y'(t) = 0$ corresponds with a normal mode of the x-oscillator (Primary System) and the question is whether nontrivial y-oscillation can reduce the oscillations of the Primary System. The system is rather artificial but contains many interesting phenomena that are typical for autoparametric interaction.

System (3.13) is invariant under $(x, y) \rightarrow (x, -y)$, $(x, y) \rightarrow (-x, y)$, and $(x, y) \rightarrow (-x, -y)$.

Using averaging we can investigate the stability of solutions of system (3.13) following [23] in more detail. Introduce the transformation leading to variational systems suitable for non-autonomous systems (1.16):

$$\begin{cases} x = u_1 \cos t + v_1 \sin t & ; \quad x' = -u_1 \sin t + v_1 \cos t \\ y = u_2 \cos t + v_2 \sin t & ; \quad y' = -u_2 \sin t + v_2 \cos t. \end{cases}$$

(3.14)

For simpler notation we rescale $t = \frac{\varepsilon}{2}\tilde{t}$. The averaged system of system (3.13) becomes:

$$u_1' = -k_1 u_1 + (\sigma_1 - \frac{1}{2}a)v_1 + v_1(u_1^2 + v_1^2) + \frac{1}{4}c_1 u_2^2 v_1 + \frac{3}{4}c_1 v_2^2 v_1 + \frac{1}{2}c_1 u_2 v_2 u_1$$

$$v_1' = -k_1 v_1 - (\sigma_1 + \frac{1}{2}a)u_1 - u_1(u_1^2 + v_1^2) - \frac{3}{4}c_1 u_2^2 u_1 - \frac{1}{4}c_1 v_2^2 u_1 - \frac{1}{2}c_1 u_2 v_2 v_1$$

$$u_2' = -k_2 u_2 + \sigma_2 v_2 + v_2(u_2^2 + v_2^2) + \frac{1}{4}c_2 u_1^2 v_2 + \frac{3}{4}c_2 v_1^2 v_2 + \frac{1}{2}c_2 u_1 v_1 u_2$$

$$v_2' = -k_2 v_2 - \sigma_2 u_2 - u_2(u_2^2 + v_2^2) - \frac{3}{4}c_2 u_1^2 u_2 - \frac{1}{4}c_2 v_1^2 u_2 - \frac{1}{2}c_2 u_1 v_1 v_2.$$

(3.15)

Although there are many parameters involved this system can be analysed for critical points, periodic and quasi-periodic solutions, producing existence of periodic solutions and detailed stability diagrams in parameter space, see [23]. A criterion for energy absorption by the y-oscillator is that the normal mode x-oscillator (the Primary System) is unstable. In system (3.15) this normal mode corresponds with $u_2 = v_2 = 0$. To find periodic solutions we have to compute critical points of the equations for u_1, v_1. It turns out that to find a critical point $u_1 = \alpha, v_1 = \beta$ we have the condition $a > 2k_1$. The stability of the normal mode $(u_1, v_1, u_2, v_2) = (\alpha, \beta, 0, 0)$ is obtained by computing the Jacobian of the righthand side of system (3.15) and determining the eigenvalues in the critical point.

It turns out there exist open sets of parameter values of damping, detuning and parametric excitation where the normal mode is unstable. This enables quenching of the x-oscillator by autoparametric coupling to the y-oscillator.

System (3.13) has also been analysed by the numerical bifurcation program CONTENT, a routine followed up later by MATCONT. Interestingly, it is shown in [23] that the system contains a sequence of period-doubling bifurcations leading to chaotic solutions.

3.3   The Poincaré-Lindstedt Method

We discussed the Poincaré-Lindstedt method briefly in Sect. 1.8. This method was developed for periodic solutions and uses a periodicity condition that makes it related to averaging. At the same time this focus on periodic solutions excludes results on more general orbits and stability. A strong point is that the Poincaré-Lindstedt method leads to a convergent series of approximations whereas averaging results are, apart from periodic solutions, asymptotic.

A basic ingredient of the method is the Poincaré expansion theorem:

Theorem 3.2 *Consider the initial value problem*

$$\dot{x} = F(t, x, \varepsilon), x(t_0) = \mu, \tag{3.16}$$

with $|t - t_0| < h$, $x, \mu \in D \subset \mathbb{R}^n$, $0 \leq \varepsilon \leq \varepsilon_0$. If $F(t, x, \varepsilon)$ is continuous with respect to t, x and ε and can be expanded in a convergent power series with respect to x and ε for $||x|| \leq \rho, 0 \leq \varepsilon \leq \varepsilon_0$ (ρ a positive constant), then $x(t)$) can be expanded in a convergent power series with respect to ε and μ in a neighbourhood of $\varepsilon = 0$,

$$x(t) = x_0(t) + \varepsilon x_1(t) + \ldots + \varepsilon^n x_n(t) + \ldots$$

convergent on the timescale 1.

Although the timescale of validity for the expansion is $O(1)$ and so relatively short, it is sufficient for periodic solutions with $O(1)$ period. The proof of theorem (3.2) was given by Poincaré in the 1st volume of [50], a modern proof is given in [76]. The title of the 3 volumes [50] is deceptive, although a large number of applications in the books are on celestial mechanics the books can be considered as the first systematic text on nonlinear dynamics. See also the discussion in [78].

The procedure of Poincaré-Lindstedt is based on theorem (3.2) and the Picard-Lindelöf method of iteration; this iteration is actually based on contraction of maps applied to ODEs that satisfy a Lipschitz condition as formulated in the Introduction. See for more details of the ODE background [20] chs. 1 and 2. It consists of writing the ODE with initial value as an integral equation. Substituting a suitable guess for the solution (as we saw in the Introduction even using the initial value is sufficient) the iteration process converges to the exact solution. However the convergence is slow and many integrations over polynomials and other functions are involved. So the process of iteration is correct but without additional assumptions impractical. The contraction is directly related to the implicit function theorem and the iteration process can be drastically improved by using Poincaré expansion and imposing special conditions. For periodic solutions we expand and apply periodicity conditions.

To demonstrate the procedure we start with the variational equation (3.1) that we studied by averaging.

Example 3.3 We write the Eq. (3.1) as an integral equation:

$$x(t) = x_0 + \varepsilon \int_0^t (x(s) - 2x(s)^2 \sin^2(s))ds, \tag{3.17}$$

where we have chosen $t_0 = 0$; as for $\varepsilon = 0$ we have $\dot{x} = 0$, we choose for the unknown initial value $x(0)$ the expansion:

$$x(0) = \rho_0 + \varepsilon\rho_1 + \varepsilon^2\rho_2 + \varepsilon^3 \dots$$

As the equation is 2π-periodic we will look for 2π-periodic solutions. Substuting the expansion into the integral equation (3.17) we find with high order terms omitted:

$$x(0) + \varepsilon x_1(t) + \varepsilon^2 x_2(t) + \varepsilon^3 \dots =$$

$$x(0) + \varepsilon \int_0^t [(x(0) + \varepsilon x_1(s) + \varepsilon^2 x_2(s) - 2(x(0) + \varepsilon x_1(s) + \varepsilon^2 x_2(s))^2 \sin^2 s]ds.$$

Using the expansion for $x(0)$ and arranging by order of ε we find to first order:

$$x_1(t) = \int_0^t [\rho_0 - 2\rho_0^2 \sin^2 s]ds = (\rho_0 - \rho_0^2)t + \frac{1}{2}\rho_0^2 \sin 2t.$$

For the periodic solution one has the *periodicity condition*:

$$\int_0^{2\pi} x_1(s)ds = 0 \Rightarrow \rho_0 - \rho_0^2 = 0.$$

We conclude that $\rho_0 = 0$ or $\rho_0 = 1$. Choose $\rho_0 = 1$ as earlier for Eq. (3.1).
Keeping the $O(\varepsilon)$ terms in the integral we have $x_2(t) =$

$$\int_0^t [\rho_1 + x_1(s) - 2(2\rho_0\rho_1 + 2\rho_0 x_1(s)) \sin 2s]ds$$

$$= \int_0^t [\rho_1 + \frac{1}{2} \sin 2s - 4(\rho_1 + \frac{1}{2} \sin 2s)^2 \sin 2s]ds.$$

Integration leads to the secular term for $x_2(t)$: $-\rho_1 t$ so $\rho_1 = 0$. We conclude for the periodic solution:

$$x(t) = 1 + \frac{1}{2}\varepsilon \sin 2t + O(\varepsilon^2). \qquad (3.18)$$

We compare with the exact solution (3.2) by expanding:

$$x_s(t) = \frac{4 + \varepsilon^2}{4 + \varepsilon^2 - 2\varepsilon \sin 2t - \varepsilon^2 \cos 2t} = 1 + \frac{1}{2}\varepsilon \sin 2t + O(\varepsilon^2), \ x(0) = 1 + \frac{1}{4}\varepsilon^2.$$

3.3.1 General Approach for Variational Systems

Generalising to n-dimensional variational systems runs in the same way. Consider a system as derived from a perturbation problem as in the introductory Chap. 1:

$$\dot{x} = \varepsilon f(t, x, \varepsilon), \qquad (3.19)$$

that satisfies the conditions of theorem (10.39) and with the vector field $f(t, x)$ T-periodic in t. We write down the corresponding integral equation

$$x(t) = x(0) + \int_0^t f(s, x(s), \varepsilon) ds,$$

we expand the vector field $f(t, x)$ with respect to ε, substitute the Poincaré-expansion for $x(t)$ and expand the unknown initial condition with respect to ε in constant n-vectors $x(0) = \rho_0, \rho_1, \ldots$ Equating powers of ε we find integral equations for the terms $x_n(t), n = 0, 1, 2, \ldots$ like (3.17). If we find by the periodicity condition a unique, isolated solution ρ_0 for $x_1(t)$ to be periodic, we have pinpointed the periodic solution to first order. If we have no uniqueness we can try to find this at next order. This is tied in with the implicit function theorem as will be made clear in the next subsection.

Using the Poincaré-Lindstedt method for autonomous systems we will in general have a perturbation of the period T suggested by the original perturbation problem. Using as a start the corresponding variational system in amplitude-phase coordinates this has as advantage a time dependent phase that may perturb the unperturbed period T.

In some cases, as for Hamiltonian systems, we have families of periodic solutions and we find no uniqueness at any order of ε. In such cases we have to modify the Poincaré-Lindstedt method.

3.3.2   The Poincaré-Lindstedt Approach for Autonomous Systems

As we discussed, the iteration process using the integral equation form of an ODE with initial value is correct and produces results on a timescale of $O(1)$. However, the effectiveness of the iteration depends on the type of solution we are looking for and on the ODE. For instance in the case of Eq. (3.19) we are looking for T-periodic solutions. Consider now the 2nd order autonomous equation:

$$\ddot{x} + x = \varepsilon f_1(x, \dot{x}) + \varepsilon^2 f_2(x, \dot{x}) + O(\varepsilon^3) \tag{3.20}$$

and assume that the assumptions of the Poincaré expansion theorem (3.2) are satisfied. A periodic solution, if it exists, may have a period close to the 2π-periodic solutions of the unperturbed equation, but its actual period is still unknown. We will start with ignoring the presence of variational equations, in the applications we return to this. Following Poincaré we will look for T-periodic solutions with $T = T(\varepsilon) = 2\pi + \varepsilon \dots$. We introduce a new timelike variable θ by the near-identity transformation:

$$\omega t = \theta, \omega^{-2} = 1 - \varepsilon \eta(\varepsilon) = 1 - \eta_0 \varepsilon - \eta_1 \varepsilon^2 + O(\varepsilon^3),$$

with the purpose of looking for 2π-periodic solutions in θ. The initial values are:

$$x(0) = \rho_0 + \varepsilon \rho_1 + O(\varepsilon^2), \dot{x}(0) = 0.$$

The assumption $\dot{x}(0) = 0$ is no restriction for 2nd order autonomous ODEs. Equation (3.20) becomes with these transformations:

$$\frac{d^2x}{d\theta^2} + x = \varepsilon \eta x + \varepsilon(1 - \varepsilon \eta) f(x, (1 - \varepsilon \eta)^{-1/2} \frac{dx}{d\theta}) + \varepsilon^2 \dots \tag{3.21}$$

Consider the integral equation that is equivalent to the initial value problem on θ:

$$x(\theta) = x(0) \cos \theta + \varepsilon$$

$$\int_0^\theta \sin(\theta - s)[\eta x(s) + (1 - \varepsilon \eta) f_1(x(s), (1 - \varepsilon \eta)^{-1/2} \frac{dx}{d\theta})] ds + \varepsilon^2 \dots \tag{3.22}$$

where the $O(\varepsilon^2)$ terms are determined by $f_1, f_2, \dots$. For $\varepsilon = 0$ Eq. (3.20) has an infinite set of periodic, synchronous (harmonic) functions. We will look for periodic solutions in θ that branch off of one or more harmonic functions in this set. The periodicity condition for the first term f_1 becomes:

$$\int_0^{2\pi} sin(\theta - s)[\eta x(s) + (1 - \varepsilon\eta)f_1(x(s), (1 - \varepsilon\eta)^{-1/2}\frac{dx}{d\theta})]ds = 0. \quad (3.23)$$

Writing $sin(\theta - s) = \sin\theta\cos s - \cos\theta\sin s$ we find from the integral 2 conditions:

$$\begin{cases} F_1 &= \int_0^{2\pi} \cos s[\eta x(s) + (1 - \varepsilon\eta)f_1(x(s), (1 - \varepsilon\eta)^{-1/2}\frac{dx}{d\theta})]ds = 0, \\ F_2 &= \int_0^{2\pi} \sin s[\eta x(s) + (1 - \varepsilon\eta)f_1(x(s), (1 - \varepsilon\eta)^{-1/2}\frac{dx}{d\theta})]ds = 0. \end{cases}$$
$$(3.24)$$

To lowest order we have 2 unknowns, ρ_0 and η_0, at each order of ε again 2 unknowns. At each step the system (3.24) is uniquely solvable if for the Jacobian $\Delta = det(J) \neq 0$:

$$J = \frac{\partial(F_1, F_2)}{\partial(\rho_n, \eta_n)}, n = 0, 1, 2\ldots \quad (3.25)$$

If $det(J)$ does not vanish, the periodicity condition satisfies the implicit function theorem.

To lowest order we have the conditions:

$$\begin{cases} F_1 &= \pi\eta_0\rho_0 + \int_0^{2\pi} \cos s \, f_1(\rho_0\cos s, -\rho_0\sin s)ds = 0, \\ F_2 &= \int_0^{2\pi} \sin s \, f_1(\rho_0\cos s, -\rho_0\sin s)ds = 0. \end{cases} \quad (3.26)$$

If the lowest order system (3.26) satisfies already the uniqueness condition we have existence of the periodic solution corresponding with ρ_o, η_0.

Remark 3.4 In the spirit of Lindstedt we have started with one perturbed harmonic equation. The extension to systems of autonomous equations is straightforward.

3.3.3 The Poincaré-Lindstedt Approach for Non-autonomous Systems

A 2nd order equation like (3.20) with T-periodic solutions for $\varepsilon = 0$ requires an expansion of time to describe periodic solutions with period close to T. Also, the initial condition will in general depend on ε. We have 2 parameter expansions.

This changes if we have an equation with T-periodic terms like:

$$\ddot{x} + x = \varepsilon f(t, x, \dot{x}, \varepsilon) \quad (3.27)$$

that satisfies the Poincaré-expansion theorem (3.2) and with $f(t, x, \dot{x}, \varepsilon)$ T-periodic. A periodic solution, if it exists, will have period T. For autonomous equations the transformation for time and phase ϕ: $t + \phi = \tau$ leaves the equations invariant, but this is not the case for non-autonomous equations. So we introduce the phase ϕ as a parameter to be determined. The initial value provides the 2nd parameter for Eq. (3.27).

An alternative is to use the transformation (1.16) that introduces implicitly a phase. The set-up of the Poincaré-Lindstedt method, the use of the integral equation, periodicity conditions and application of the implicit function theorem runs as before. In [76, ch. 10.3] this is shown for the forced Duffing equation with damping:

$$\ddot{x} + \varepsilon \mu \dot{x} + x - \varepsilon x^3 = \varepsilon h \cos \omega t,$$

see also Sect. 3.2.3 for the treatment by averaging. We will see another application, the Mathieu-equation with damping, in the sequel.

3.4   Applications of the Poincaré-Lindstedt Method

Although the Van der Pol-equation has been extensively discussed in various forms and generalisation, also in many books, we will discuss the equation again because of its iconic character. The equation was also used often in papers demonstrating fomula-manipulating computer routines; see for an early example [2]. Other typical examples illustrating the method will follow.

3.4.1   The Periodic Solution of the Van der Pol-Equation

We can directly copy from the treatment of autonomous equations in the case:

$$\ddot{x} + x = \varepsilon \dot{x}(1 - x^2).$$

Equation (3.21) becomes:

$$\frac{d^2x}{d\theta^2} + x = \varepsilon \eta x + \varepsilon(1 - \varepsilon\eta)^{1/2}(1 - x^2)\frac{dx}{d\theta}.$$

The corresponding integral equation is:

$$x(\theta) = x(0) \cos \theta + \varepsilon \int_0^\theta \sin(\theta - s)[\eta x(s) + (1 - \varepsilon\eta)^{1/2}(1 - x(s)^2)\frac{dx(s)}{ds}]ds.$$

Comparing with (3.26) we have the periodicity conditions:

$$F_1 = \pi \eta_0 \rho_0 = 0,$$

$$F_2 = \frac{1}{2}\rho_0(1 - \frac{1}{4}\rho_0^2) = 0.$$

The periodicity conditions yield $\rho_0 = 2$, $\eta_0 = 0$.

It is not difficult to implement the iteration process for automatic computation. In [2] this was done for 164 steps.

Using the Variational Equations

As noted in Sect. 3.3.2 we can use the variational equations (2.13):

$$\dot{r} = \varepsilon r \sin^2(t + \phi)(1 - r^2 \cos^2(t + \phi)),$$

$$\dot{\phi} = \varepsilon \sin(t + \phi) \cos(t + \phi)(1 - r^2 \cos^2(t + \phi)).$$

The integral equations are:

$$\begin{cases} r(t) &= r(0) + \varepsilon \int_0^t r(s) \sin^2(s + \phi(s)) \left(1 - r^2(s) \cos^2(s + \phi(s))\right) ds, \\ \phi(t) &= \phi(0) + \varepsilon \int_0^t \sin(s + \phi(s)) \cos(s + \phi(s)) \left(1 - r^2(s) \cos^2(s + \phi(s))\right) ds. \end{cases}$$

$$(3.28)$$

As stated earlier putting $\phi_0 = 0$ is no restriction for autonomous second order equations. To obtain only periodic terms we require the secular terms for the amplitude to vanish. This demonstrates that the Poincaré-Lindstedt method is complementary to averaging as instead of removing the terms with average zero, we remove the secular terms.

Using the expansion theorem we have $\phi(t) = 0 + \varepsilon \phi_1 + \ldots, r(t) = r_0 + \varepsilon r_1 + \ldots$ with the dots standing for higher order terms dependent on time, so to $O(\varepsilon)$ we start with $r_0 \cos t$. The first integral equation produces:

$$\int_0^t \frac{1}{2} r_0 \left(1 - \frac{1}{4} r_0^2 - \cos 2s + \frac{1}{4} r_0^2 \cos 4s\right) ds.$$

As expected we find the periodic secularity condition $r_0 = 2$ and to first order in ε:

$$r(t) = 2 + \varepsilon(-\frac{1}{2} \sin 2t + \frac{1}{4} \sin 4t) + O(\varepsilon^2).$$

$$(3.29)$$

For $\phi(t)$ we find to first order in ε the 2π-periodic expression:

$$\phi(t) = \varepsilon(\frac{1}{4} \cos 2t + \frac{1}{8} \cos 4t) + O(\varepsilon^2)$$

$$(3.30)$$

Including the $O(\varepsilon)$ terms the solution is still 2π-periodic, a period change will be at most $O(\varepsilon^2)$.

3.4.2   A Case of Non-uniqueness

In some cases where the Jacobian vanishes this is caused by the presence of a continuous famlly of periodic solutions. Additional information may help to identify such cases. In Example 1.4 we discussed a simple Hamiltonian system with a continuous family of periodic solutions. Consider the system again.

Example 3.4 The equation from Example 1.4 for $n = 3$ is:

$$\ddot{x} + x - \varepsilon x^3 = 0.$$

In the formulation of Eq. (3.22) we have the integral equation:

$$x(\theta) = x(0)\cos\theta + \varepsilon \int_0^\theta \sin(\theta - s)[\eta x(s) + (1 - \varepsilon\eta)x^3(s)]ds.$$

To first order this leads to the periodicity conditions:

$$F_1 = \eta_0\rho_0 + \int_0^{2\pi} \cos s\, \rho_0^3 \cos^3 s\, ds = \rho_0(\eta_0 + \frac{3}{8}\rho_0^2) = 0,$$

$$F_2 = \int_0^{2\pi} \sin s\, \rho_0^3 \cos^3 s\, ds = 0.$$

The 2nd periodicity condition holds for any ρ (as expected from Example 1.4), for η_0 we find:

$$\eta_0 = -\frac{3}{8}\rho_0^2.$$

3.4.3   The Mathieu-Equation

The Mathieu-equation is an important example of parametric excitation. Several parameters are involved, we considered cases with and without damping, see Eq. (1.5). We found that in parameter-space the stable and unstable solutions are separated by curves forming a boundary of the so-called instability-tongues. On the separating curves the solutions are periodic. The Poincaré-Lindstedt method should be able to describe these curves. We will use the variational equations.

Example 3.5 In the Introduction, Example 1.4, we discussed the equation by averaging with $\omega = 1$. We reformulate as:

$$\ddot{x} + (\omega^2 + \varepsilon \cos 2t)x = 0.$$

We assume that ω is ε-close to 1 and that we have a convergent expansion of the frequency by:

$$\omega^2 = 1 - \beta_0 \varepsilon + O(\varepsilon^2)$$

The period π is present by the excitation in the equation. If $\varepsilon = 0$ we give the solution as:

$$x|_{\varepsilon=0}(t) = y_1(0) \cos t + y_2(0) \sin t, \ \dot{x}|_{\varepsilon=0}(t) = -y_1(0) \sin t + y_2(0) \cos t$$

Using the variational system with transformation (1.16) we have to first order:

$$\dot{y}_1 = -\varepsilon(\beta_0 - \cos 2t)(y_1(t) \cos t + y_2(t) \sin t) \sin t,$$
$$\dot{y}_2 = -\varepsilon(\beta_0 - \cos 2t)(y_1(t) \cos t + y_2(t) \sin t) \cos t.$$

The equivalent integral equations for $y_1(t)$, $y_2(t)$ produce the periodicity conditions:

$$\begin{cases} \int_0^{2\pi} (\beta_0 - \cos 2s(y_1(s) \cos s + y_2(s) \sin s) \sin s \, ds &= 0, \\ \int_0^{2\pi} (\beta_0 - \cos 2s(y_1(s) \cos s + y_2(s) \sin s) \cos s \, ds &= 0. \end{cases} \tag{3.31}$$

The periodicity conditions to first order are:

$$y_2(0) \left(\beta_0 + \frac{1}{2} \right) = 0, \ y_1(0) \left(\beta_0 - \frac{1}{2} \right) = 0.$$

We have the boundaries of the instability tongue $\beta_0 = \pm\frac{1}{2}$ as we found by averaging in Example 3.2.5.

A few additional examples can be found in the monographs [63] and [77].

Chapter 4
Second Order Periodic Averaging

To improve the averaging approximation to second order in ε we can be motivated by the necessity of more precision but, more importantly, also by the fact that qualitative new phenomena arise at higher order. The following theorem gives such an improvement in the case of periodic vector fields.

4.1 Second Order Precision

Theorem 4.1 *Consider the initial value problems*

$$\dot{x} = \varepsilon f(t, x) + \varepsilon^2 g(t, x) + \varepsilon^3 R(t, x, \varepsilon), x(t_0) = x_0, \tag{4.1}$$

and

$$\dot{y} = \varepsilon f^0(y) + \varepsilon^2 f_1^0(y) + \varepsilon^2 g^0(y), y(t_0) = x_0, \tag{4.2}$$

with

$$f_1(t, x) = Df(t, x)u^1(t, x) - Du^1(t, x)f^0(x),$$

and

$$u^1(t, x) = \int_{t_0}^{t_0+t} (f(s, x) - f^0(x))ds.$$

© The Author(s), under exclusive license to Springer Nature Switzerland AG 2023
F. Verhulst, *A Toolbox of Averaging Theorems*, Surveys and Tutorials in the Applied
Mathematical Sciences 12, https://doi.org/10.1007/978-3-031-34515-9_4

Assume for $x, y, x_0 \in D \subset \mathbb{R}^n$, $t_0 \leq t \leq \infty$, $0 \leq \varepsilon \ll 1$:

1. *f, g, R, Df are defined, continuous and bounded by a constant M independent of ε in $[t_0, \infty] \times D$ (the square matrix Df refers to the Jacobian derivative with respect to the spatial variable x only).*
2. *R is Lipschitz-continuous with respect to $x \in D$;*
3. *f and g are smooth and T-periodic in T with T a constant independent of ε, f^0, f_1^0, g^0 are the averages of f, f_1, g;*
4. *$y(t)$ belongs to a bounded interior subset of D on the timescale $1/\varepsilon$,*

then we have for the solutions of the initial value problems the estimate:

$$x(t) - (y(t) + \varepsilon u^1(t, y)) = O(\varepsilon^2) \text{ as } \varepsilon \to 0 \text{ on the timescale } 1/\varepsilon.$$

Remark 4.1 For a proof see [58, ch. 2.9]. As stated before a second order approximation is only required when the first order approximation produces trivial results or if we expect new phenomena at second order. This may happen more often than the literature on perturbation theory suggests. Interesting examples are the instability tongues of the Mathieu-equation and higher order resonances in Hamiltonian systems.

Remark 4.2 As $u^1(t, x)$ has average zero, in the periodic case $Du^1(t, x)f^0(x)$, will have average zero.

Remark 4.3 As in the case of first order averaging the case of quasi-periodic averaging carries over to second order. In the case of almost-periodic vector fields where we have an infinite number of independent periods the analysis is more complicated, see also Chap. 5.

4.2 Applications

The periodic solution of the Van der Pol-equation is calculated to second order in ch. 2 of [58]. The important qualitative change is that in the expression $x(t) = r(t)\cos((t + \phi(t))$ we have in second order approximation dependence on $(t - \frac{1}{16}\varepsilon^2 t)$.

An interesting question is if the periodic solution of the Van der Pol-equation (2.12) persists if the self-excitation takes place at $O(\varepsilon^2)$ and at order $O(\varepsilon)$ the oscillations are not harmonic. More applications can be found in Chaps. 6 and 8. For Hamiltonian and dissipative systems with dimensions larger than 2, qualitative phenomena will often shift to higher order. We start with some straightforward questions about 2-dimensional conservative systems.

4.2.1 Simple Conservative Systems

Consider the equation:

$$\ddot{x} + x = \varepsilon(ax^2 + bx^3). \tag{4.3}$$

Equation (4.3) can be derived form the Hamiltonian function:

$$H = \frac{1}{2}\dot{x}^2 + \frac{1}{2}x^2 - \varepsilon\frac{a}{3}x^3 - \varepsilon\frac{b}{4}x^4. \tag{4.4}$$

We are interested in the influence of the nonlinear terms in Eq. (4.3) on the radius r and phase ϕ that affects the period. Suppose $x(0) = r_0 > 0$, $\dot{x}(0) = 0$. From the Hamiltonian we have the first integral:

$$\frac{1}{2}\dot{x}^2 + \frac{1}{2}x^2 - \varepsilon\frac{a}{3}x^3 - \varepsilon\frac{b}{4}x^4 = C, \ C = \frac{1}{2}r_0^2 - \varepsilon\frac{a}{3}r_0^3 - \varepsilon\frac{b}{4}r_0^4$$

With transformation (1.6) we have $\dot{x}^2(t) + x^2(t) = r^2(t)$, so it is clear that near the origin of phase-space $r(t) = r_0 + O(\varepsilon)$ for $t \geq 0$. We will investigate with more precision what happens to radius and phase (period) in the quadratic and cubic case.

Example 4.1 Consider the **quadratic** case $a \neq 0, b = 0$. Applying transformation (1.6) with $r(0) = r_0, \phi(0) = 0$ we find:

$$f(t, r, \phi) = \begin{pmatrix} -ar^2 \sin(t + \phi)\cos^2(t + \phi) \\ -ar\cos^3(t + \phi) \end{pmatrix},$$

with

$$f^0(r, \phi) = \begin{pmatrix} 0 \\ 0 \end{pmatrix}.$$

So $r(t) = r_0 + O(\varepsilon), \phi(t) = 0 + O(\varepsilon)$. To obtain a 2nd order approximation following Theorem 4.1 we have to compute and average the expression $f_1 = Df.u^1$. Abbreviating $t + \phi = \alpha$, we find $Df(t, r, \phi) =$

$$\begin{pmatrix} -2ar\sin\alpha\cos^2\alpha & -ar^2\cos^3\alpha + 2ar^2\sin^2\alpha\cos\alpha \\ -a\cos^3\alpha & 3ar\cos^2\alpha\sin\alpha \end{pmatrix},$$

We find:

$$u^1(t, r, \phi) = \begin{pmatrix} \frac{1}{3}ar^2\cos^3\alpha \\ -ar\sin\alpha + \frac{1}{3}ar\sin^3\alpha \end{pmatrix}.$$

We find from $Df.u^1$ after averaging: $(0, -\frac{5}{8}a^2r^2)^T$.

In the notation of Theorem 4.1 we have:

$$\dot{y}_1 = 0, \quad \dot{y}_2 = -\varepsilon^2 \frac{5}{8} a^2 r^2,$$

and so the $O(\varepsilon^2)$ approximation for $r(t)$ becomes:

$$r(t) = r_0 + \varepsilon \frac{1}{3} a r_0^2 \cos^3(t - \varepsilon^2 \frac{5}{8} a^2 r_0^2 t) \tag{4.5}$$

The approximation for the phase can be obtained in the same way.

In the case of a cubic term the first order approximation is not trivial.

Example 4.2 Consider the **cubic** case $a = 0, b \neq 0$. Applying transformation (1.6) with $r(0) = r_0, \phi(0) = 0$ we find:

$$f(t, r, \phi) = \begin{pmatrix} -br^3 \sin(t + \phi) \cos^3(t + \phi) \\ -br^2 \cos^4(t + \phi) \end{pmatrix}, \quad f^0(r, \phi) = \begin{pmatrix} 0 \\ -\frac{3}{8} br^2 \end{pmatrix}.$$

Using initial values $r(0) = r_0, \phi(0) = 0$, we have a first-order approximation valid on the timescale $1/\varepsilon$:

$$x(t) = r_0 \cos(1 - \varepsilon \frac{3}{8} br_0^2)t + O(\varepsilon), \quad \dot{x}(t) = -r_0 \sin(1 - \varepsilon \frac{3}{8} br_0^2)t + O(\varepsilon),$$

corresponding with a set of periodic solutions with an $O(\varepsilon)$ shifted period that depends on the initial r_0. Following Theorem 4.1 we have to compute and average again $f_1 = Df.u^1$. Abbreviating again $t + \phi = \alpha$, we find

$$Df(t, r, \phi) = \begin{pmatrix} -3br^2 \sin \alpha \cos^3 \alpha & -br^3 \cos^4 \alpha + 3br^3 \sin^2 \alpha \cos^2 \alpha \\ -2br \cos^4 \alpha & 4br^2 \cos^3 \alpha \sin \alpha \end{pmatrix},$$

and, using $\cos^4 \alpha = \frac{3}{8} + \frac{1}{2} \cos 2\alpha + \frac{1}{8} \cos 4\alpha$,

$$u^1(t, r, \phi) = \begin{pmatrix} \frac{1}{8} br^3 \cos 2\alpha + \frac{1}{32} br^3 \cos 4\alpha \\ -\frac{1}{4} br^2 \sin 2\alpha - \frac{1}{32} br^2 \sin 4\alpha \end{pmatrix}.$$

Multiplying Df and u^1 we find f_1 and after averaging

$$f_1^0(r, \phi) = \begin{pmatrix} 0 \\ -\frac{51}{256} b^2 r^4 \end{pmatrix}.$$

In the notation of theorem (4.2) we obtain:

$$\dot{y}_1 = 0, \ \dot{y}_2 = -\varepsilon \frac{3}{8} b y_1^2 - \varepsilon^2 \frac{51}{256} b^2 y_1^4. \tag{4.6}$$

We can use this result to improve the approximation of r, ϕ but, more interesting, we will use this 2nd order approximation in the next subsection.

4.2.2 Van der Pol-Excitation at 2nd Order

We summarise a result of [77, ch. 12.5]. Consider the equation:

$$\ddot{x} + x - \varepsilon b x^3 = \varepsilon^2 \dot{x}(1 - x^2)$$

with b a suitable constant independent of ε. Introducing amplitude-phase variables with transformation (1.6) we find (see preceding example):

$$f(t, r, \phi) = \begin{pmatrix} -br^3 \sin(t + \psi) \cos^3(t + \phi) \\ -br^2 \cos^4(t + \phi) \end{pmatrix}, \ f^0(r, \phi) = \begin{pmatrix} 0 \\ -\frac{3}{8} br^2 \end{pmatrix}.$$

Using initial values $r(0) = r_0, \phi(0) = 0$, we have a first-order approximation valid on the timescale $1/\varepsilon$:

$$x(t) = r_0 \cos(1 - \varepsilon \frac{3}{8} b r_0^2)t + O(\varepsilon), \ \dot{x}(t) = -r_0 \sin(1 - \varepsilon \frac{3}{8} b r_0^2)t + O(\varepsilon),$$

corresponding with a set of periodic solutions with an $O(\varepsilon)$ shifted period that also depends on the initial r_0. This family of periodic solutions does not persist as a periodic solution branches off from one of the first-order approximate solutions when considering second-order approximations. We show this. Following Theorem 4.1 we use the result from the preceding example for $f_1 = Df.u^1$ after averaging:

$$f_1^0(r, \phi) = \begin{pmatrix} 0 \\ -\frac{51}{256} b^2 r^4 \end{pmatrix}.$$

Now we can write down the equation for y as formulated in the procedure for second-order calculations. Note that $g^0(y)$ (notation Theorem 4.1) was calculated already. We obtain

$$\dot{y}_1 = \varepsilon^2 \frac{y_1}{2} \left(1 - \frac{1}{4} y_1^2 \right)$$

$$\dot{y}_2 = -\varepsilon \frac{3}{8} b y_1^2 - \varepsilon^2 \frac{51}{256} b^2 y_1^4.$$

If $y_1(0) = r_0 = 2$, we have a stationary solution $y_1(t) = 2$, and with $y_2(0) = \phi(0) = 0$

$$y_2(t) = -\varepsilon \frac{3}{2} bt - \varepsilon^2 \frac{51}{16} b^2 t.$$

The equilibrium is obtained at $O(\varepsilon^2)$ of the averaged system and is asymptotically stable. An $O(\varepsilon^2)$-approximation of the asymptotically stable periodic solution is obtained by inserting y_1, y_2 into u^1 so that

$$r(t) = 2 + \varepsilon b \cos 2(t + y_2(t)) + \varepsilon \frac{b}{4} \cos 4(t + y_2(t)) + O(\varepsilon^2),$$

$$\phi(t) = y_2(t) - \varepsilon b \sin 2(t + y_2(t)) - \varepsilon \frac{b}{8} \sin 4(t + y_2(t)) + O(\varepsilon^2),$$

valid on the timescale $1/\varepsilon$. At this order of approximation, the timelike variables for the periodic solution are $t, \varepsilon t, \varepsilon^2 t$.

4.2.3 Intermezzo on the Nature of Timescales

As discussed earlier, an assumption is often made in research that the natural timelike variables in expansions with respect to a small parameter ε will be of the form $t, \varepsilon t, \varepsilon^2 t, \ldots, \varepsilon^n t$ etc. As we discussed in the Introduction this assumption is usually based on simple examples and it is sometimes correct but in general not. For instance in Chaps. 7 and 8 we will meet systems where local domains play a part with completely different timelike variables.

In [77] (section 12.5) and [79] a practical problem, the Mathieu-equation, is studied in the form:

$$\ddot{x} + (1 + \varepsilon a + \varepsilon^2 b + \varepsilon \cos 2t) x = 0. \tag{4.7}$$

In [77] a first order approximation produces the boundaries of the instability tongue or Floquet tongue sketched in Fig. 3.1 (without damping) with $a = \pm 0.5$. The $O(\varepsilon)$ approximation is valid on the interval of time $1/\varepsilon$. A 2nd order calculation shows that the boundary of the tongue is pinpointed with $b = -1/32$; the timelike variables that play a part at 2nd order are $t, \varepsilon t, \varepsilon^{3/2} t, \varepsilon^2 t$. This is quite unexpected and this result would not have been obtained with apriori assumptions on the timelike variables.

The unexpected timelike variables arise in a neighbourhood of transitions like the curves of periodic solutions of the Mathieu-equation that separate domains of stability and instability and are in general associated with bifurcations. *Such phenomena are often of special interest in nonlinear analysis as they correspond with qualitative changes in the system.* As an illustration we discuss a few examples involving low dimensional bifurcations (nvolving few variables and parameters); we can find such bifurcations as subsystems after first or second order averaging.

A simple example is the saddle-node bifurcation described by:

$$\dot{x} = a - bx^2. \tag{4.8}$$

If $ab < 0$ there is no critical point; suppose $ab > 0$, then we have the critical points $x_0 = \pm\sqrt{a/b}$. Their stability and local behaviour with time is described by the coefficient of linearisation $-2bx_0 = \mp 2\sqrt{ab}$ near the critical points. If for instance $a = \varepsilon^2, b = \varepsilon$ the leading timelike variable is $\varepsilon^{3/2}t$.

Consider the system inspired by the pitchfork bifurcation:

$$\dot{x} = \varepsilon^2 y - \varepsilon y^3, \quad \dot{y} = \varepsilon x. \tag{4.9}$$

Three critical points (equilibria) are $(x_0, y_0) = (0, 0), (0, \pm\sqrt{\varepsilon})$. Linearisation near the critical points $(0, y_0)$ produces:

$$\dot{x} = \varepsilon^2 y - 3\varepsilon y_0^2 y + \dots, \quad \dot{y} = \varepsilon x,$$

where the dots represent the neglected nonlinear terms. We have the characteristic eigenvalue equations and timelike variables near the critical points:

(0, 0): $\lambda^2 - \varepsilon^3 = 0$, timelike variable $\varepsilon^{3/2}t$.

$(0, \pm\sqrt{\varepsilon})$: $\lambda^2 - 2\varepsilon^3 = 0$, timelike variable $\varepsilon^{3/2}t$.

See for extensive discussions of bifurcations and timelike variables also [44]. In general we expect in regions where bifurcations occur the presence of timelike variables of the form $\varepsilon^q t$ with q a positive rational number.

As we will see, an example of the pitchfork bifurcation is found in the Van der Pol-equation.

Example 4.3 (Emergence of the Van der Pol Periodic Solution)

Consider the Van der Pol-equation in slightly more general form than Eq. (2.12):

$$\ddot{x} + x = \varepsilon\dot{x}(a - x^2). \tag{4.10}$$

If parameter $a < 0$, the oscillations will be damped, there is no periodic solution. If $a > 0$ we have nearly the case of Eq. (2.12), first order averaging produces:

$$\dot{r} = \frac{\varepsilon}{2}r(a - \frac{1}{4}r^2), \quad \dot{\phi} = 0. \tag{4.11}$$

If parameter a starts at a negative value and we let it increase it will pass through zero and for $a > 0$ a periodic solution emerges with amplitude $2\sqrt{a}$ by a pitchfork bifurcation. It starts with a being small, say $a = \varepsilon$. This is the situation sketched above where we have the timelike variable $\varepsilon^{3/2}t$. We improve the reasoning by writing Eq. (4.10) as:

$$\ddot{x} + x = -\varepsilon \dot{x} x^2 + \varepsilon^2 \dot{x}.$$

In amplitude-phase variables r, ϕ the variational system to $O(\varepsilon)$ becomes:

$$\dot{r} = -\varepsilon r^3 \sin^2(t + \phi) \cos^2(t + \phi),$$
$$\dot{\phi} = -\varepsilon r^2 \sin(t + \phi) \cos^2(t + \phi).$$

First order averaging produces:

$$\dot{r} = -\varepsilon \frac{1}{8} r^3, \ \dot{\phi} = 0.$$

Computing the quantities Df and u^1 in the notation of Theorem 4.1 we find no $O(\varepsilon^2)$ contribution for the amplitude and a small contribution to the phase. Averaging the term $\varepsilon^2 \dot{x}$ we find:

$$\dot{r} = -\varepsilon \frac{1}{8} r^3 + \varepsilon^2 \frac{r}{2}. \tag{4.12}$$

As predicted above we find for the amplitude of the periodic solution $2\sqrt{\varepsilon}$.

4.2.4 The Chaotic Systems Sprott A and NE8

We discuss the Sprott A system (NE1) and NE8 system that are cases of 17 chaotic, 3-dimensional systems, $NE1, \ldots, NE17$, with linear and quadratic terms only; they were listed and studied in [38]. We follow [1] to consider the systems near the origin of phase-space where chaos is not dominant.

Example 4.4 (Sprott A (or NE1))

$$\dot{x} = y, \ \dot{y} = -x - \varepsilon yz, \ \dot{z} = \varepsilon y^2 - \varepsilon a, \tag{4.13}$$

with positive parameter a. Note that the system has no critical points (equilibria), but it has an invariant manifold given by $x = y = 0$, the z-axis with solutions, $z(t) = z(0) - \varepsilon a t$. Transforming to slowly varying cylindrical variables r, ϕ, z we find:

$$\begin{cases} \dot{r} &= -\varepsilon rz \sin^2(t + \phi), \\ \dot{\phi} &= -\frac{1}{2} \varepsilon z \sin(2t + 2\phi), \\ \dot{z} &= \varepsilon r^2 \sin^2(t + \phi) - \varepsilon a. \end{cases} \tag{4.14}$$

Averaging while keeping the same notation for variables we have for the components of f^0:

$$\dot{r} = -\frac{1}{2}\varepsilon rz, \ \dot{\phi} = 0, \ \dot{z} = \varepsilon(\frac{1}{2}r^2 - a). \tag{4.15}$$

An equilibrium is $(r, \phi, z) = (\sqrt{2a}, \phi(0), 0)$. As expected for autonomous systems, see Remark 3.3, one of the eigenvalues of the Jacobian Df^0 is zero; the rank of the Jacobian at equilibrium is 2. We conclude with Theorem 3.1 that a periodic solution exists in an $O(\varepsilon)$-neighbourhood of the equilibrium. However, the eigenvalues of the Jacobian at equilibrium are $\pm ai, 0$. In the case of purely imaginary eigenvalues the nonlinear terms may stabilise or destabilise the equilibrium. We call this a structurally unstable case. It turns out that a second order approximation does not add real terms $O(\varepsilon^2)$ to the eigenvalues, so the stability of this periodic solution remains unsolved by averaging.

It was shown in [9] that because of symmetries in the Sprott A system we have a special situation. The periodic solution is surrounded by an infinite family of tori and is neutrally stable.

Example 4.5 (System NE8) A related case with a different outcome is another system of the list in [38], system NE8:

$$\dot{x} = y, \ \dot{y} = -x - \varepsilon yz, \ \dot{z} = \varepsilon(xy + \frac{1}{2}x^2 - a,) \tag{4.16}$$

with positive parameter a. Note that again this system has no critical points (equilibria), but it has an invariant manifold given by $x = y = 0$, the z-axis with solutions, $z(t) = z(0) - \varepsilon at$. Transforming to slowly varying cylindrical variables r, ϕ, z we find:

$$\begin{cases} \dot{r} &= -\varepsilon rz \sin^2(t + \phi), \\ \dot{\phi} &= -\frac{1}{2}\varepsilon z \sin(2t + 2\phi), \\ \dot{z} &= -\frac{\varepsilon}{2}r^2 \sin(2t + 2\phi) + \frac{\varepsilon}{2}r^2 \cos^2(t + \phi) - \varepsilon a. \end{cases} \tag{4.17}$$

Averaging while keeping the same notation for variables we have for the components of f^0:

$$\dot{r} = -\frac{1}{2}\varepsilon rz, \ \dot{\phi} = 0, \ \dot{z} = \varepsilon(\frac{1}{4}r^2 - a). \tag{4.18}$$

An equilibrium is $(r, \phi, z) = (2\sqrt{a}, \phi(0), 0)$. As before one of the eigenvalues of the Jacobian Df^0 is zero; the rank of the Jacobian at equilibrium is 2. We conclude with Theorem 3.1 that a periodic solution exists in an $O(\varepsilon)$-neighbourhood of the equilibrium. The eigenvalues of the Jacobian at equilibrium are $\pm ai, 0$, so for stability we have to compute a second order approximation of the eigenvalues.

To reduce the calculations we use that $\theta = t + \phi$ can be used as a timelike variable. System (4.17) reduces to second order:

$$\begin{cases} \frac{dr}{d\theta} = -\varepsilon r z \sin^2 \theta + \varepsilon^2 g_1(\theta, r, z) + \varepsilon^3 \dots \\ \frac{dz}{d\theta} = -\frac{\varepsilon}{2} r^2 \sin 2\theta + \frac{\varepsilon}{2} r^2 \cos^2 \theta - \varepsilon a + \varepsilon^2 g_2(\theta, r, z) + \varepsilon^3 \dots \end{cases} \quad (4.19)$$

Averaging the $O(\varepsilon^2)$ terms we find $g_1^0 = 0$, $g_2^0 = -\frac{1}{8} r^2 z$. In the notation of Theorem 4.1 we find:

$$Df(\theta, r, z) = \begin{pmatrix} -z \sin^2 \theta & -r \sin^2 \theta \\ -r \sin 2\theta + r \cos^2 \theta & 0 \end{pmatrix},$$

$$u^1(\theta, r, z) = \begin{pmatrix} \frac{1}{4} r z \sin 2\theta \\ \frac{1}{4} r^2 (\cos 2\theta + \frac{1}{2} \sin 2\theta) \end{pmatrix}.$$

$$f_1^0(r, z) = \left(\frac{1}{16} r^3, -\frac{1}{4} r^2 z \right).$$

For the approximating solutions to second order we have:

$$\frac{dr}{d\theta} = -\frac{1}{2} \varepsilon r z + \varepsilon^2 \frac{1}{16} r^3, \quad \frac{dz}{d\theta} = \varepsilon (\frac{1}{4} r^2 - a) - \frac{\varepsilon^2}{4} r^2 z. \quad (4.20)$$

The equilibrium to second order becomes $(r, z) = (2\sqrt{a}, \frac{\varepsilon}{2} a)$. From Eq. (4.20) we compute the eigenvalues of this equilibrium to find that $-\frac{1}{8} \varepsilon a$ is added to the purely imaginary parts. At second order we have a qualitative change, the periodic solution is stabilised. Numerical calculations confirm the result.

Chapter 5
First Order General Averaging

We can apply averaging in a more general sense with as important examples equations with terms that have a limit as $t \to \infty$ and if almost-periodic vector fields are present. Consider again Eq. (2.1) and assume that the following 'general' average exists:

$$f^0(x) = \lim_{t \to \infty} \frac{1}{T} \int_{t_0}^{t_0+T} f(t, x)dt. \tag{5.1}$$

As we shall see general averages of the form (5.1) play a part in certain applications. It may also generate a different order function that arises in the estimates. We have with $x \in D$ an error $O(\delta(\varepsilon))$ on the timescale $1/\varepsilon$ with:

$$\delta(\varepsilon) = \sup_{x \in D} \sup_{0 \le \varepsilon t \le L} \varepsilon \left| \int_{t_0}^{t_0+t} (f(s, x) - f^0(x))ds \right|, \tag{5.2}$$

with L a positive constant.

5.1 The Basic Theorem for General Averaging

Theorem 5.1 *Consider the initial value problem* (2.1) *(brief version $\dot{x} = \varepsilon f + \varepsilon^2 g$, $x(t_0) = x_0$) and the initial value problem*

$$\dot{y} = \varepsilon f^0(y), \ y(t_0) = x_0, \tag{5.3}$$

© The Author(s), under exclusive license to Springer Nature Switzerland AG 2023
F. Verhulst, *A Toolbox of Averaging Theorems*, Surveys and Tutorials in the Applied
Mathematical Sciences 12, https://doi.org/10.1007/978-3-031-34515-9_5

with $f^0(x)$ existing and defined by Eq. (5.1). Assume for $x, y, x_0 \in D \subset \mathbb{R}^n, t_0 \leq t \leq \infty, 0 \leq \varepsilon \ll 1$:

1. f, g, Df are defined, continuous and bounded by a constant M independent of ε in $[t_0, \infty] \times D$;
2. g is Lipschitz-continuous with respect to $x \in D$;
3. $y(t)$ belongs to a bounded interior subset of D on the timescale $1/\varepsilon$,

then we have for the solutions of the initial value problems the estimate:

$$x(t) - y(t) = O(\delta(\varepsilon)) \text{ as } \varepsilon \to 0 \text{ on the timescale} 1/\varepsilon.$$

Remark 5.1 The proof is rather subtle, see [58] ch. 4. A early proof was given by Bogoliubov and Mitropolsky in [13].

The more general version of averaging is highly relevant for the treatment of so-called almost-periodic functions. We will not give an explicit definition, but see [14]. We note that for almost-periodic functions the average f^0 in the sense of Theorem 5.3 always exists. It is also important to know that there exists an analogue of Fourier theory so that a vector field $f(t, x)$ that is almost-periodic in t, can be written as a uniformly convergent series:

$$f(t, x) = \sum_{n=0}^{\infty} (A_n(x) \cos \lambda_n t + B_n(x) \sin \lambda_n t). \tag{5.4}$$

If $f(t, x)$ is periodic, we would have $\lambda_n = n$, but in this more general case the λ_n can be any sequence of real numbers. The collection of these numbers is called the generalised Fourier spectrum. As stated above for an almost-periodic vector function, the following average exists:

$$f^0(x) = \lim_{T \to \infty} \frac{1}{T} \int_0^T f(t, x) dt.$$

A strong contrast with periodic Fourier theory is that if the average $f^0(x)$ vanishes, *the primitive of $f(t, x)$ need not be bounded for all time.* This weakens the error estimate, as we shall see in an application.

Remark 5.2 A useful lemma was proved in [58]. Suppose that there exists a positive number α such that we have for the spectrum of the generalised Fourier expansion (5.4) of $f(t, x) \lambda_n \geq \alpha, n = 0, 1, 2, \ldots$, then the primitive of $f(t, x) - f^0(x)$ is bounded and the error by first order general averaging is $O(\varepsilon)$.

Remark 5.3 If the expansion (5.4) is finite we call $f(t, x)$ a quasi-periodic function. The error by first order averaging is $O(\varepsilon)$ in this case. See also Theorem 2.2.

5.2 Second Order General Averaging

Consider the slowly varying system (2.1) and assume that the general average (5.1) exists. In Theorem 5.1 we have assumed that Df is continuous. It is easy to formulate a second order approximation result if the vector field $f(t, x)$ is even smoother, at least $C^2(D)$. We transform:

$$x(t) = y(t) + \delta(\varepsilon)u^1(t, y(t)),$$

with u^1 defined as for periodic averaging. Then we can construct for a large number of systems, dependent on the choice of $f(t, x)$, an $O(\delta^2)$ approximation of $x(t)$ on time intervals of size $1/\varepsilon$.

That we can not obtain a second order approximation in some cases depends on the rate of convergence of the limit (5.1). For details and proofs see lemma 4.5.1 in [58]. We will discuss some aspects of second order general averaging in subsequent applications.

5.3 Applications

In physics and engineering the conditions for general averaging are quite natural. Still, the number of applications is restricted. If we are considering equations with time-dependent terms that are not periodic and even show limiting behaviour we may come across the concept of *adiabatic invariant*. This invariant plays usually the part of approximate time-dependent integral of motion in such dissipative equations.

5.3.1 Van der Pol-Equation with Changing Friction

Consider the equation:

$$\ddot{x} + x = \varepsilon(1 - ae^{-t} - x^2)\dot{x}, \ a > 0. \tag{5.5}$$

Transforming to amplitude-phase with (1.6) we find the variational equations:

$$\dot{r} = \varepsilon \sin(t + \phi)(1 - ae^{-t} - r^2\cos^2(t + \phi))r\sin(t + \phi),$$

$$\dot{\phi} = \frac{\varepsilon}{r}\cos(t + \phi)(1 - ae^{-t} - r^2\cos^2(t + \phi))r\sin(t + \phi).$$

We have

$$0 < \int_0^t ae^{-s}\sin^2(s + \phi)ds \leq \int_0^t ae^{-s}ds = a(1 - e^{-t}).$$

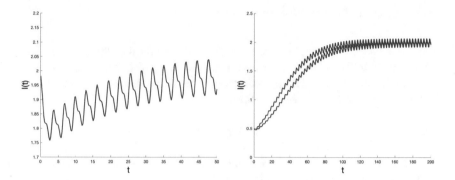

Fig. 5.1 Behaviour of $I = 0.5(\dot{x}^2 + x^2)$ with time of the Van der Pol-equation, with changing damping in Eq. (5.5), $\varepsilon = 0.05$, $a = 2$, $x(0) = 0$, $\dot{x}(0) = 2$. Right the start at $x(0) = 0$, $\dot{x}(0) = 1$ and both exponential and polynomial variation of the damping

We conclude that

$$\lim_{t \to \infty} \frac{1}{T} \int_0^T ae^{-s} \sin^2(s + \phi)ds = 0.$$

A similar estimate holds for the equation for ϕ. We conclude that the estimate for the approximations of the standard Van der Pol-equation carry over to this case with changing damping, see Fig. 5.1 where initially the action $I(t)$ moves away from the end stage near $I = 2$. For comparison we show in Fig. 5.1 (right) the times series starting at $x(0) = 0$, $\dot{x}(0) = 1$ with $-a^{-t}$ in the equation and this term replaced by $-a/(1 + t)$ moving slower to $I = 2$.

5.3.2 Linear Oscillations with Increasing Friction, Adiabatic Invariant

Consider a linear oscillator with friction that, through wear and tear, increases slowly with time. The equation is

$$\ddot{x} + \varepsilon f(t)\dot{x} + x = 0. \tag{5.6}$$

Assume that $f(0) = 1$ and $f(t)$ increases monotonically with time. We discuss 2 examples:

$$f_1(t) = 2 - e^{-t^2}, \quad f_2(t) = 2 - \frac{t}{1 + t^2}.$$

In amplitude-phase variables r, ψ, we have with transformation (1.6):

$$\dot{r} = -\varepsilon r \sin^2(t + \phi) f(t),$$
$$\dot{\psi} = -\varepsilon \sin(t + \phi) \cos(t + \phi) f(t).$$

We can perform averaging in the more general sense, and for both choices $f = f_1(t)$ and $f = f_2(t)$ we find the approximating system:

$$\dot{r} = -\varepsilon r, \quad \dot{\psi} = 0.$$

In both cases, we have on the timescale $1/\varepsilon$ the approximate solution:

$$x(t) = r(0)e^{-\varepsilon t} \cos(t + \psi(0)) + O(\delta(\varepsilon)). \tag{5.7}$$

However, the error is different. For the function $f_1(t)$, the integral $\int_0^t (f(s, x) - f^0(x))ds$ is uniformly bounded, so the error is $O(\varepsilon)$. Choosing $f_2(t)$, the integral is logarithmically growing as $\ln(1 + t^2)$; we have on the timescale $1/\varepsilon$: $\delta(\varepsilon) = O(\varepsilon |\ln \varepsilon|)$.

Such a difference is not unexpected. Note that the approximation corresponds with a constant friction coefficient 2, which is the limit for $t \to \infty$ of $f_1(t)$ and $f_2(t)$. The function $f_1(t)$ reaches this limit exponentially fast, but $f_2(t)$ increases polynomially to the value 2.

The non-autonomous equation (5.6) has the time-varying amplitude

$$r(t) = r(0)e^{-\varepsilon t}$$

as approximate quantity; this is called an *adiabatic invariant* of the non-autonomous equation (5.6).

If the variation in time of the friction is even slower, for instance by taking

$$f(t) = 2 - \frac{1}{(1 + t)^\alpha}, 0 < \alpha < 1,$$

then a first order approximation exists with an $O(\varepsilon^\alpha)$ approximation, but a second order approximation with the construction mentioned above does not exist. See also the related example 4.5.1 in [58].

5.3.3 Quasi-Periodic Forcing of the Duffing-Equation

Although we can handle many quasi-periodic problems with the basic techniques of Chap. 2, the behaviour of quasi-periodic systems is of special interest as it represents relative simple almost-periodic problems. Consider the forced Duffing-quation:

$$\ddot{x} + x + \varepsilon \gamma x^3 = \varepsilon f(t, x), \tag{5.8}$$

with $\gamma \neq 0$ a constant, $f(t, x)$ is almost-periodic in t. Transforming with (1.6) we find:

$$\dot{r} = \varepsilon \sin(t + \phi)(\gamma r^3 \cos^3(t + \phi) - f(t, x)),$$

$$\dot{\phi} = \varepsilon \frac{\cos(t+\phi)}{r}(\gamma r^3 \cos^3(t + \phi) - f(t, x))).$$

Assume that $f(t, x)$ is quasi-periodic of the form:

$$f(t, x) = \sum_{n=1}^{N}(A_n(x) \cos(\lambda_n t) + B_n(x) \sin \lambda_n t), \tag{5.9}$$

with independent (incommensurable) frequencies λ_n.

As a first example we take for $f(t, x)$: $f_1(t, x) = x(\cos t + \cos \sqrt{2}t)$. Applying the basic quasi-periodic averaging Theorem 2.2 (Remark 5.9) we find for the averaged equations:

$$\dot{r} = 0, \quad \dot{\psi} = \frac{3}{8}\varepsilon\gamma r^2. \tag{5.10}$$

We conclude that on the timescale $1/\varepsilon$ the amplitude is constant with error $O(\varepsilon)$, the phase ψ changes slowly. A numerical approximation confirms this analysis for a long time.

Consider an almost-periodic function of the form:

$$f_2(t, x) = \sum_{n=0}^{\infty} \frac{x}{(2n + 1)^2} \sin\left(\frac{t}{2n + 1}\right). \tag{5.11}$$

The spectrum accumulates near zero. The approximating system (5.10) will be the same but the error will be larger than $O(\varepsilon)$. This is a consequence of the fact that the integral over time of the function $f_2(t, x)$ is not bounded.

But in the case of for instance the choice:

$$f_3(t, x) = \sum_{n=1}^{\infty} \frac{x}{(2n + 1)^2} \sin\left(\sqrt{n}t\right)$$

we would have the same approximation with error $O(\varepsilon)$ (see Remark 5.2, $\alpha = 1$).

5.3.4 Quasi-Periodic Forcing of a Van der Pol Limit Cycle

Consider the Van der Pol-equation (2.12) but with $O(1)$ forcing added. A simple case is:

$$\ddot{x} + x = \varepsilon\dot{x}(1 - x^2) + f(t), \tag{5.12}$$

with $f(t)$ a quasi-periodic function; put:

$$f(t) = \sum_{n=1}^{N} a_n \cos \omega_n t, \tag{5.13}$$

with N a natural number, the frequencies $\omega_n, n = 1, \dots, N$ are irrational numbers, incommensurable and not ε-close to 1.

If we replace $f(t)$ in Eq. (5.12) by $\varepsilon f(t)$ first order averaging produces the result of the standard Van der Pol-equation except that we have no existence of a periodic solution. We conjecture for ε small the existence of a quasi-periodic solution moving on a torus in the with time extended phase-space; see Fig. 5.2.

Consider $f(t) = O(1), t \geq 0$. If $\varepsilon = 0$ a particular solution $\eta(t)$ of Eq. (5.12) is:

$$\eta(t) = \sum_{n=1}^{N} \frac{a_n}{1 - \omega_n^2} \cos \omega_n t, \tag{5.14}$$

Transforming with (1.27) for forced equations we put:

$$x = r \cos(t + \phi) + \eta(t), \quad \dot{x} = -r \sin(t + \phi) + \dot{\eta}(t).$$

to find for r, ψ:

$$\dot{r} = -\varepsilon \sin(t + \phi)(-r \sin(t + \phi) + \dot{\eta}(t))(1 - (r \cos(t + \phi) + \eta(t))^2),$$

$$\dot{\phi} = -\frac{\varepsilon}{r} \cos(t + \phi)(-r \sin(t + \phi) + \dot{\eta}(t)) \left(1 - (r \cos(t + \phi) + \eta(t))^2 \right).$$

Fig. 5.2 Projection on the $(x_1, x_2) = (x, \dot{x})$ phase-plane for the forced Van der Pol-equation (5.12) in the case of small forcing $\varepsilon \cos \sqrt{2} t$ with $\varepsilon = 0.05$

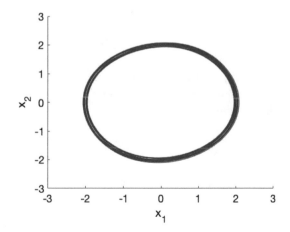

The righthand side is quasi-periodic in t; as an example we choose:

$$f(t) = a \cos \sqrt{\omega} t. \tag{5.15}$$

Averaging produces the system:

$$\dot{r} = \frac{\varepsilon}{2} r (1 - \frac{1}{4} r^2 - \frac{1}{2} \frac{a^2}{(1 - \omega^2)^2}), \ \dot{\phi} = 0.$$

The forcing causes an important change of the standard Van der Pol-equation. If $\omega^2 = 2$ the approximate solution with constant r vanishes if $1 - a^2/2$ becomes negative so if $a \geq \sqrt{2}$. Experiments confirm that in such cases the forcing dominates and produces quasi-periodic solutions. In various parameter cases the dynamics of the forced Van der Pol-equation can be quite complicated.

5.3.5   Evolution to Symmetry, Adiabatic Invariants

This subsection is based on parts of [83]. The paper uses the observation that many macroscopic structures in nature are characterised by symmetries. One can think of planets and stars that are usual spherical and when rotating are flattened at the poles, globular clusters of stars, the final stage of tidal evolution in the gravitational two-body problem and many types of galaxies. To describe the evolution of these systems with time in detail using physical laws is very complicated, so we choose a shortcut to evolution by assuming that we have initially a non-symmetric potential whereas the asymmetries vanish with time. We start with a cartoon problem.

Example 5.1 Consider a 2-dimensional system governed by a potential with asymmetric term vanishing:

$$\Phi(x) = \frac{1}{2} x^2 + \frac{\varepsilon}{3} f(t) x^3, \tag{5.16}$$

with $f(0) = 1$ and $f(t)$ a continuous function with the property $f(t) \to 0$ as $t \to \infty$ monotonically. For instance $f(t) = \exp(-t)$ or $f(t) = 1/(1+t)$. Consider the equation:

$$\ddot{x} + x = \frac{\varepsilon a}{1 + t} x^2. \tag{5.17}$$

If $\varepsilon = 0$ or parameter $a = 0$, the action I

$$I = \frac{1}{2} (\dot{x}^2 + x^2) \tag{5.18}$$

will be constant, $I = 0.5(\dot{x}^2(0) + x^2(0))$. The nonlinear term will vanish with time and the question is whether the original dynamical state has still some influence after a long time. We use transformation (1.6) to obtain the variational equations:

$$\dot{r} = -\varepsilon \sin(t + \phi)\frac{a}{1+t}r^2 \cos^2(t + \phi), \quad \dot{\phi} = -\varepsilon \cos(t + \phi)\frac{a}{1+t}r \cos^2(t + \phi).$$

Using $\sin(t + \phi) \cos^2(t + \phi) = \frac{1}{4}(\sin(t + \phi) + \sin(3t + 3\phi))$ and Theorem 5.1 we have to evaluate the integral for $\dot{r}$ (leaving out constants):

$$f_1^0(x) = \lim_{t\to\infty} \frac{1}{T}\int_0^T \frac{1}{1+t}(\sin(t + \phi) + \sin(3t + 3\phi))dt.$$

Using partial integration the integral becomes before taking the limit:

$$\cos\phi(0) - \frac{\cos(t + \phi)}{1+t} + \cos 3\phi(0) - \frac{\cos(3t + 3\phi)}{3(1+t)}$$

$$- \int_0^t \frac{1}{(1+t)^2}(\cos(t + \phi) + \frac{1}{3}\cos(3t + 3\phi))dt$$

The integral converges for $t \to \infty$ so we find the average $\dot{r} = 0$. In the same way we find the averaged equation $\dot{\phi} = 0$. We conclude that we have with $r(0) = r_0$, $\phi(0) = \phi_0$ the approximation:

$$x(t) = r_0 \cos(t + \phi_0) + \delta(\varepsilon) \text{ on the timescale } 1/\varepsilon. \tag{5.19}$$

The amplitude is an adiabatic invariant of Eq. (5.17). We can estimate $\delta(\varepsilon) = \varepsilon|\ln\varepsilon|$. An improved first order estimate can be obtained by using convergent iteration of the corresponding integral equation:

$$x(t) = r_0 \cos(t + \phi_0) + \varepsilon a \int_0^t \sin(t - s)\frac{x^2(s)}{1+s}ds.$$

Putting $\phi_0 = 0$, $x(s) = r_0 \cos s$ we find after partial integration as a first step:

$$\tilde{x}(t) = r_0 \cos t + \varepsilon\frac{a}{4}r_0^2\left(\sin t\,(3\frac{\sin t}{1+t} + \frac{\sin 3t}{3(1+t)} + \cos t\,(\frac{\cos t}{1+t} + \frac{\cos 3t}{3(1+t)} - \frac{4}{3})\right).$$

The convergence of the expansion is in t, not in ε. This is why we call $\tilde{x}(t)$ an improved first order approximation. It is clear from the approximation that the initial values play a part in the later dynamics; see also Fig. 5.3.

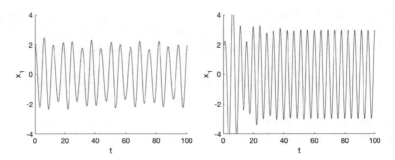

Fig. 5.3 Two timeseries $x_1(t)$ for the forced Van der Pol-equation (5.12) with $f(t) = a \cos \sqrt{2}t$. Left the case $a = 0.3$ where $r(t)$ does not tend to zero and is modulated by the quasi-periodic forcing. Right the case $a = 3$ where $r(t)$ vanishes and the motion $x(t)$ tends to the particular solution $-3 \cos \sqrt{2}t$

The evolution results can be more spectacular if the evolutionary dynamics changes the stability of solutions. This is for 1 dof not so easy to achieve but it may happen for 2 dof as we will show for a special case of the Hamiltonian:

$$H = \frac{1}{2}(\dot{x}^2 + x^2 + \dot{z}^2 + \omega^2 z^2) - \varepsilon(\frac{1}{3}a_1 x^3 + a_2 x z^2 + e^{-\delta t}\frac{1}{3}a_3 z^3 + e^{-\delta t}a_4 x^2 z).$$
(5.20)

The parameter $\delta > 0$ will govern the dynamics of evolution, the terms asymmetric in z will vanish with time. We consider a case studied in [83] and keep the notation of that paper to make the connection easier. In [83] we describe the motion of stars in a galactic plane with $x(t)$ de epicyclic motion in the plane with respect to circular orbits, $z(t)$ the vertical oscillation with respect to the galactic plane. Consider the case $a_1 = a_3 = 0$, $\omega = 2$ producing the 2 dof system:

$$\begin{cases} \ddot{x} + x &= \varepsilon a_2 z^2 + \varepsilon e^{-\delta t} 2a_4 xz, \\ \ddot{z} + 4z &= \varepsilon 2a_2 xz + \varepsilon e^{-\delta t} a_4 x^2. \end{cases}$$
(5.21)

We observe that the x normal mode ($z = \dot{z} = 0$) exists and is harmonic, the z normal mode does not exist (Fig. 5.4).

Example 5.2 Choosing $\delta = 1$ we have relatively fast evolution in the 2 dof system (5.21). General averaging produces as in the preceding example no contribution of the asymmetric terms on the timescale $1/\varepsilon$, so the dynamical effects are small. Putting $\delta = \varepsilon$ will produce less trivial results.

We transform to slowly varying polar coordinates r, ϕ by (1.6):

$$x = r_1 \cos(t + \phi_1), \dot{x} = -r_1 \sin(t + \phi_1), z = r_2 \cos(2t + \phi_2), \dot{z} = -2r_2 \sin(2t + \phi_2),$$

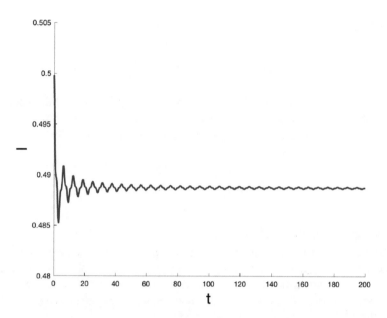

Fig. 5.4 Time series for the action $I(t)$ of Eq. (5.17) with $\varepsilon = 0.05, a = 1, x(0) = 1, \dot{x}(0) = 0$. We have $I(0) = 0.5$ and $I(t)$ tending to 0.488

leading to the slowly varying system:

$$
\begin{cases}
\dot{r}_1 &= -\varepsilon \sin(t + \phi_1)\Big(a_2 r_2^2 \cos^2(2t + \phi_2) \\
&\quad -e^{-\varepsilon t} 2a_4 r_1 \cos(t + \phi_1) r_2 \cos(\omega t + \phi_2)\Big), \\
\dot{\phi}_1 &= -\varepsilon \frac{\cos(t+\phi_1)}{r_1}\Big(a_2 r_2^2 \cos^2(2t + \phi_2) \\
&\quad -e^{-\varepsilon t} 2a_4 r_1 \cos(t + \phi_1) r_2 \cos(2t + \phi_2)\Big), \\
\dot{r}_2 &= -\frac{\varepsilon}{2} \sin(2t + \phi_2)\Big(2a_2 r_1 \cos(t + \phi_1) r_2 \cos(2t + \phi_2) \\
&\quad -e^{-\varepsilon t} a_4 r_1^2 \cos^2(t + \phi_1)\Big), \\
\dot{\psi}_2 &= -\varepsilon \frac{\cos(2t+\phi_2)}{2r_2}\Big(2a_2 r_1 \cos(t + \phi_1) r_2 \cos(2t + \phi_2) \\
&\quad -e^{-\varepsilon t} \sin(2t + \phi_2) a_4 r_1^2 \cos^2(t + \phi_1)\Big).
\end{cases}
\tag{5.22}
$$

Near the normal modes $r_1 = 0$ and $r_2 = 0$ we have to use a different coordinate transformation.

We can put $\tau = \varepsilon t$ and treat τ as a new, independent variable. We add the equation:

$$
\dot{\tau} = \varepsilon.
$$

It will be useful to introduce the actions E_1, E_2 by:

$$E_x = \frac{1}{2}(\dot{x}^2 + x^2) = \frac{1}{2}r_1^2, \ E_z = \frac{1}{2}(\dot{z}^2 + \omega^2 z^2) = \frac{\omega^2}{2}r_2^2. \tag{5.23}$$

First order averaging produces that in system (5.22) the terms with coefficients a_2 vanish. The implication is *that to first order in ε and on time intervals of size $1/\varepsilon$* system (5.21) is described by the intermediate normal form equations:

$$\begin{cases} \ddot{x} + x & = \varepsilon e^{-\tau} 2a_4 xz, \\ \ddot{z} + \omega^2 z & = \varepsilon e^{-\tau} a_4 x^2, \\ \dot{\tau} & = \varepsilon. \end{cases} \tag{5.24}$$

The Relation with Dissipation
As the nonlinear terms are homogeneous in the coordinates we can remove the time-dependent term by a transformation involving $e^{-\tau}$. We transform $x = e^{-\tau} y_1, z = e^{-\tau} y_2$ Such a time-dependent transformation exists for any positive sufficiently differentiable function of time that decreases monotonically to zero replacing $e^{-\varepsilon t}$ in system (5.21). System (5.24) transforms to:

$$\begin{cases} \ddot{y}_1 + y_1 & = -2\varepsilon \dot{y}_1 - \varepsilon^2 y_1 + \varepsilon 2a_4 y_1 y_2, \\ \ddot{y}_2 + 4y_2 & = -2\varepsilon \dot{y}_2 - \varepsilon^2 y_2 + \varepsilon a_4 y_1^2. \end{cases} \tag{5.25}$$

So the time-dependent transformation transforms the asymmetric system to an equivalent dissipative system with friction coefficient 2ε. We can study the resulting dissipative system but we return to the original formulation.

We will average system (5.24) using transformation (1.6) keeping $r_1, r_2, \phi_1, \phi_2, \tau$ fixed. We find to first order with combination angle $\chi = 2\phi_1 - \phi_2$:

$$\begin{cases} \dot{r}_1 & = -\varepsilon e^{-\tau} \frac{a_4}{2} r_1 r_2 \sin \chi, \ \dot{\phi}_1 = -\varepsilon e^{-\tau} \frac{a_4}{2} r_2 \cos \chi, \\ \dot{r}_2 & = \varepsilon e^{-\tau} \frac{a_4}{8} r_1^2 \sin \chi, \ \dot{\phi}_2 = -\varepsilon e^{-\tau} \frac{a_4}{8} \frac{r_1^2}{r_2} \cos \chi, \end{cases} \tag{5.26}$$

For χ we have the equation:

$$\frac{d\chi}{dt} = \varepsilon a_4 e^{-\tau}(-r_2 + \frac{r_1^2}{8r_2}) \cos \chi. \tag{5.27}$$

It is surprising that to this first order system (5.26) admits a time-independent integral of motion (adiabatic invariant):

$$\frac{1}{2}r_1^2 + 2r_2^2 = E_0, \tag{5.28}$$

with constant $E_0 \geq 0$; In the original coordinates we have:

$$\frac{1}{2}(\dot{x}^2 + x^2) + \frac{1}{2}(\dot{z}^2 + 4z^2) = E_0.$$

We will see that the separate actions E_x, E_z are not conserved. However, system (5.26) admits special families of solutions with constant amplitude on the timescale $O(1/\varepsilon)$ if:

$$\chi = 0, \pi, r_1^2 = 8r_2^2, \ r_1^2 = \frac{4}{3}E_0, r_2^2 = \frac{1}{6}E_0. \tag{5.29}$$

The solutions with $\chi = 0$ are called in-phase, the solutions with $\chi = \pi$ out-phase. The corresponding phases are slowly decreasing with rate $\exp(-\varepsilon t)$. These special solutions are found in a so-called resonance manifold, a geometric structure that will be studied in more detail in Chap. 8.

The solutions of system (5.26) have the error estimate $O(\varepsilon)$ on the timescale $1/\varepsilon$. On this long interval of time we expect the terms $O(\varepsilon^2)$ to play a part as the solutions of system (5.26) with coefficients a_4 will vanish and maybe terms with coefficient a_2 arise.

Second Order Averaging
To improve the approximations by second order averaging [58] of system (5.22), we borrow the result from [83]:

$$\begin{cases} \dot{r}_1 = -\varepsilon e^{-\tau} \frac{a_4}{2} r_1 r_2 \sin\chi, \ \dot{r}_2 = \varepsilon e^{-\tau} \frac{a_4}{8} r_1^2 \sin\chi, \\ \dot{\phi}_1 = -\varepsilon e^{-\tau} \frac{a_4}{2} r_2 \cos\chi - \varepsilon^2 e^{-2\tau} \frac{1}{64} a_4^2 (9r_1^2 + 4r_2^2), \\ \dot{\phi}_2 = -\varepsilon e^{-\tau} \frac{a_4}{8} \frac{r_1^2}{r_2} \cos\chi - \varepsilon^2 \left(\frac{1}{30} a_2^2 r_1^2 + \frac{29}{120} a_2^2 r_2^2 + e^{-2\tau} \frac{1}{32} a_4^2 r_1^2 \right). \end{cases} \tag{5.30}$$

We do not find $O(\varepsilon^2)$ terms for the amplitudes, such terms arise only for the angles. For the combination angle χ we find:

$$\begin{cases} \frac{d\chi}{dt} = \varepsilon a_4 e^{-\tau} (-r_2 + \frac{r_1^2}{8r_2}) \cos\chi + \\ \varepsilon^2 a_2^2 \left(\frac{1}{30} r_1^2 + \frac{29}{120} r_2^2 \right) + \varepsilon^2 a_4^2 e^{-2\tau} \left(-\frac{1}{32}(9r_1^2 + 4r_2^2) + \frac{1}{32} r_1^2 \right). \end{cases} \tag{5.31}$$

We find, using Eq. (5.29), that in the resonance manifold we have that $\dot{\chi} \neq 0$. So the solutions with constant amplitude do not persist at second order.

1. During an interval of time of order $1/\varepsilon$ the integral (adiabatic invariant) (5.28) is active, the system is still dominated by the asymmetric a_4 term.
2. When time goes on the system (5.21) is dominated by the a_2 term. In [64] it is shown that for this system corresponding with a time-independent Hamiltonian, depending on the value of a_2, 2 resonance manifolds can exist on the energy manifold (see also Chap. 8).

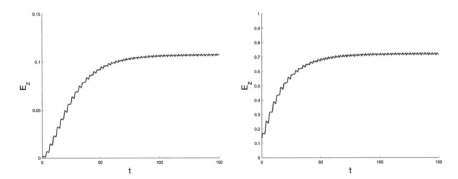

Fig. 5.5 The behaviour of action $E_z(t)$ of system (5.21) in 2 cases with $a_2 = 1$, $a_4 = 2$, $\varepsilon = 0.05$. Starting near the x normal mode (left) with $x(0) = 1$, $\dot{x}(0) = 0$, $z = 0.01$, $\dot{z}(0) = 0$. And right near an unstable solution in the resonance manifold with $x(0) = -\sqrt{2}$, $\dot{x}(0) = 0$, $z(0) = 0$, $\dot{z}(0) = 0.5$. In both cases the dynamics has changed considerably by the evolution to mirror symmetry

3. An interesting consequence arises when considering orbits that change stability during the process of evolution to mirror symmetry. The instability of some periodic solutions with constant amplitude and some normal modes persists at second order but leads to stability in the final stage of mirror symmetry after a time longer than $1/\varepsilon$. The implication is that evolution to symmetry will generally produce a drastic change of the distribution of the actions. Another qualitative consequence in this example is that the epicyclic x-normal mode does not exist in the first stage (time order $1/\varepsilon$) but the normal mode exists in the final stage of mirror symmetry; again we can expect action changes. As discussed above similar considerations hold for the unstable out-phase periodic solutions derived before (Fig. 5.5).

Chapter 6
Approximations on Timescales Longer than $1/\varepsilon$

For nonlinear variational systems of the form $\dot{x} = \varepsilon f(t, x)$ we have apriori existence of solutions on the timescale $1/\varepsilon$. With special assumptions it is possible to extend the timescale of existence and of approximations. We will formulate theorems in the case of 2 rather different sets of assumptions.

6.1 When First Order Periodic Averaging Produces Trivial Results

If in the case of periodic averaging (notation Chap. 2) we have $f^0(x) = 0$ it follows that with $x(0) = x_0$, $|x(t) - x_0| = O(\varepsilon)$ on the timescale $1/\varepsilon$. It is possible that significant qualitative phenomena take place on longer intervals of time. If $f^0(x) = 0$ we can improve the approximation and extend the timescale of validity. We formulate the theorem.

Theorem 6.1 *Consider the initial value problems*

$$\dot{x} = \varepsilon f(t, x) + \varepsilon^2 g(t, x) + \varepsilon^3 R(t, x, \varepsilon), \, x(t_0) = x_0, \tag{6.1}$$

with $f^0(x) = 0$ and

$$\dot{y} = \varepsilon^2 f_1^0(y) + \varepsilon^2 g^0(y), \, y(t_0) = x_0, \tag{6.2}$$

with

$$f_1(t, x) = Df(t, x).u^1(t, x) - Du^1(t, x).f^0(x),$$

and

$$u^1(t, x)) = \int_{t_0}^{t_0+t} f(s, x)ds.$$

Assume for $x, y, x_0 \in D \subset \mathbb{R}^n, t_0 \leq t \leq \infty, 0 \leq \varepsilon \ll 1$:

1. *f, g, R, Df are defined, continuous and bounded by a constant M independent of ε in $[t_0, \infty] \times D$;*
2. *R is Lipschitz-continuous with respect to $x \in D$;*
3. *f and g are smooth and T-periodic in T with T a constant independent of ε, f^0, f_1^0, g^0 are the averages of f, f_1, g;*
4. *$y(t)$ belongs to a bounded interior subset of D on the timescale $1/\varepsilon$,*

then we have for the solutions of the initial value problems the estimate:

$$x(t) - (y(t) + \varepsilon u^1(t, y)) = O(\varepsilon) \text{ as } \varepsilon \to 0 \text{ on the timescale } 1/\varepsilon^2.$$

Remark 6.1 Theorem 6.1 was formulated and proved by Van der Burgh in [71], see also [72] and [58] ch. 2.9.

6.2   The Case of Attraction

Another type of extension of timescales happens when attraction is involved. We present a theorem in a simple form. Essential is that we have attraction in linear approximation to an equilibrium or critical (fixed) point.

Theorem 6.2 *Consider the initial value problem (2.1) (short: $\dot{x} = \varepsilon f + \varepsilon^2 g$):*

$$\dot{y} = \varepsilon f^0(y), \quad y(t_0) = x_0. \tag{6.3}$$

Assume for $x, y, x_0, a \in D \subset \mathbb{R}^n, t_0 \leq t \leq \infty, 0 \leq \varepsilon \ll 1$:

1. *f, g, Df are defined, continuous and bounded by a constant M independent of ε in $[t_0, \infty] \times D$;*
2. *g is Lipschitz-continuous with respect to $x \in D$;*
3. *f is T-periodic in T with T a constant independent of ε;*
4. *$y(t)$ belongs to a bounded interior subset of D on the timescale $1/\varepsilon$,*
5. *$f^0(a) = 0$ with $x = a$ an asymptotically stable point in linear approximation of Eq. (6.3).*
6. *x_0 is in an interior subset of the domain of attraction of $x = a$.*

then we have for the solutions of the initial value problems the estimate:

$$x(t) - y(t) = O(\varepsilon) \text{ as } \varepsilon \to 0 \text{ for } t \geq t_0.$$

Remark 6.2 Extensions of the theorem are useful in the case of limit cycles of autonomous systems and in dynamical systems described by maps with attracting manifolds. For a proof of Theorem 6.2 and discussion see [58] ch. 5.

6.3 Applications

As mentioned before interesting new phenomena can take place at higher order approximation and on longer timescales.

6.3.1 Excitation Frequency $\omega = 1$ for the Mathieu-Equation

Consider the Mathieu-equation with a different excitation frequency:

$$\ddot{x} + (1 + 2\varepsilon \cos \omega t)x = 0, \quad \omega = 1. \tag{6.4}$$

In slowly-varying phase-amplitude variables we find with the transformation (1.6):

$$\begin{cases} \dot{r} = \frac{\varepsilon}{2}r[\sin(3t + 2\phi) + \sin(t + 2\phi)], \\ \dot{\phi} = \varepsilon[\cos t + \frac{1}{2}\cos(3t + 2\phi) + \frac{1}{2}\cos(t + 2\phi)]. \end{cases} \tag{6.5}$$

In general, averaging over the quasi-periodic righthand side associated with Eq. (6.4) for which ω is not close to 2 produces zero. We construct a second order approximation to look for less trivial results. Using the notation of Theorem 5.1 we have for Df, u^1, f_1^0:

$$Df(t, r, \phi) = \begin{pmatrix} \frac{1}{2}[\sin(3t + 2\phi) + \sin(t + 2\phi)] & r[\cos(3t + 2\phi) + \cos(t + 2\phi)] \\ 0 & -\sin(3t + 2\phi) - \sin(t + 2\phi) \end{pmatrix},$$

and,

$$u^1(t, r, \phi) = \begin{pmatrix} -r[\frac{1}{6}\cos(3t + 2\phi) + \frac{1}{2}\cos(t + 2\phi)] \\ \sin t + \frac{1}{6}\sin(3t + 2\phi) + \frac{1}{2}\sin(t + 2\phi) \end{pmatrix}.$$

Averaging the product of matrix Df and vector u^1 produces a system for $\tilde{r}, \tilde{\phi}$ representing y (formulation Theorem 6.1) in expression (6.2):

$$f_1^0(\tilde{r}, \tilde{\phi}) = \begin{pmatrix} -\frac{\tilde{r}}{2}\sin 2\tilde{\phi} \\ -\frac{1}{3} - \frac{1}{2}\cos 2\tilde{\phi} \end{pmatrix}.$$

An $O(\varepsilon)$ approximation valid on the timescale $1/\varepsilon^2$ can be obtained from the system:

$$\frac{d\tilde{r}}{dt} = -\varepsilon^2 \frac{\tilde{r}}{2} \sin 2\tilde{\phi}, \quad \frac{d\tilde{\phi}}{dt} = \varepsilon^2 (\frac{1}{3} + \frac{1}{2} \cos 2\tilde{\phi}). \tag{6.6}$$

Solving system (6.6) this approximation is the transpose of:

$$(\tilde{r}, \tilde{\phi}) + \varepsilon u^1(t, \tilde{r}, \tilde{\phi}). \tag{6.7}$$

Remark 6.3 The formulation to slowly varying variables by transformation (1.6) excludes a neighbourhood of $r = 0$, but the solutions of system (6.6) suggest instability. Using transformation (1.16) we can show that also in the case of the Mathieu-equation with $\omega = 1$ the origin is unstable.

6.3.2 A Cubic Hamiltonian System in 1 : 1 Resonance

The results in this subsection are based on [74]. Consider the two degrees-of-freedom Hamiltonian system generated by the cubic Hamiltonian: (2.25) but with $\omega = 1$:

$$H(p, q) = \frac{1}{2}(p_1^2 + q_1^2) + \frac{1}{2}(p_2^2 + q_2^2) - \varepsilon q_1 q_2^2.$$

The equations of motion are:

$$\begin{cases} \ddot{q}_1 + q_1 = \varepsilon q_2^2, \\ \ddot{q}_2 + q_2 = 2\varepsilon q_1 q_2. \end{cases} \tag{6.8}$$

In Sect. 8.4.5 we will consider a related cubic Hamiltonian producing the so-called Hénon-Heiles system. Note that for system (6.8), as in application Sect. 2.3.8, putting $q_2 = \dot{q}_2 = 0$ produces a family of periodic solutions (harmonic functions), parameterised by the energy. This is the q_1-normal mode family. As in Chap. 3 slowly varying equations arise from transformation (1.6) ($q_i, \dot{q}_i \mapsto r_i, \phi_i, i = 1, 2$). We can average over time t but this produces (not very interesting) zero righthand sides. Again, to avoid too many indices and tildes we use with some abuse of notation the original variables for the approximate quantities. We will apply Theorem 6.1, so we have to compute $f_1 = Df.u^1$ (Theorem 6.1 notation). We omit the long computations involving vectorfields of dimension 4 and 4×4

matrices. Averaging of $f_1(t, r_1, \phi_1, r_2, \phi_2)$ over time yields with combination angle $\chi = 2(\phi_1 - \phi_2)$:

$$\begin{cases} \dot{r}_1 = -\frac{\varepsilon^2}{2} r_1 r_2^2 \sin \chi, \quad \dot{\psi}_1 = -\varepsilon^2 r_2^2 (\frac{1}{3} + \frac{1}{2} \cos \chi), \\ \dot{r}_2 = \frac{\varepsilon^2}{2} r_1^2 r_2 \sin \chi, \quad \dot{\psi}_2 = -\varepsilon^2 (\frac{1}{3} r_1^2 + \frac{5}{12} r_2^2 + \frac{1}{2} r_1^2 \cos \chi). \end{cases} \quad (6.9)$$

As in application Sect. 2.3.8 we have only one combination angle χ in the averaged equations. From system (6.9) the equation for angle χ is:

$$\dot{\chi} = \varepsilon^2 \left(\frac{2}{3} r_1^2 + \frac{1}{6} r_2^2 + (r_1^2 - r_2^2) \cos \chi \right). \quad (6.10)$$

Invariant Manifolds
System (6.9) has the quadratic first integral

$$r_1^2 + r_2^2 = 2E_0, \quad (6.11)$$

with E_0 a positive constant. The family of integrals (6.11) approximates the family of energy manifolds around the origin of phase-space. As the energy manifold is bounded in the neighbourhood of the origin, the approximate integral (6.11) is valid for all time with error $O(\varepsilon^2)$.

There exists a second integral of the averaged system (6.9) :

$$r_1^2 \left(\frac{1}{4} r_1^2 + \frac{1}{3} E_0 - r_2^2 \cos \chi \right) = I, \quad (6.12)$$

with integration constant determined by the initial conditions.

The integrals (6.11) and (6.12) describe together approximate families of tori embedded in the energy manifolds. The tori surround the stable periodic solutions listed below.

Periodic Solutions
It turns out we find 6 families of periodic solutions. There are 2 normal mode families, the exact q_1 normal mode family and the ε-close q_2 normal mode family. For the second family we have to change the averaging scheme by using transformation (1.16); we omit these calculations.

As in the case of application Sect. 2.3.8 we obtain short-periodic solutions by considering the cases where $\sin \chi = 0$. The term "short" indicates usually that the frequencies of these periodic solutions are close to the basic frequencies of the linearised system near the origin. We have the possibilities:

- In-phase periodic solutions $\chi = 0, 2\pi$. Using the integrals and Eq. (6.10) we find $r_1^2 = \frac{2}{3} E_0, r_2^2 = \frac{4}{3} E_0$. The periodic solutions are stable.
- Out-phase periodic solutions $\chi = \pi, 3\pi$. Using the integrals and Eq. (6.10) we find $r_1^2 = \frac{2}{3} E_0, r_2^2 = \frac{4}{3} E_0$. The periodic solutions are stable.

It is interesting that in this problem we can obtain some general position periodic solutions as exact solutions. We use as a guess $q_1 = aq_2$ with factor a to be determined. We find:

$$q_1 = \pm \frac{1}{\sqrt{2}} q_2.$$

Substitution in system (6.8) produces 2 identical equations of the form:

$$\ddot{q}_2 + q_2 = \pm \varepsilon \sqrt{2} q_2^2. \tag{6.13}$$

The solutions of Eq. (6.13) are near the origin of phase-space periodic elliptic functions, they are parameterised by the initial conditions.

The in- and out-phase periodic solutions were obtained by localising near the origin using parameter ε. The exact periodic solutions exist for arbitrary values of the parameter ε. They represent continuous families of periodic solutions connecting the origin of phase space with other critical points of the energy manifold.

6.3.3   The Amplitude of Periodic Solutions of Autonomous Equations

We noted in remark (3.3) that for periodic solutions of autonomous ODEs we will always find at least one eigenvalue zero (Lyapunov exponent 1). For the Van der Pol-equation (2.12) and the modified versions this means that the approximations obtained imply asymptotic stability but that Theorem 6.2 can not be applied. However, the attraction takes place to the orbits in linear approximation. The consequence is that the amplitude(s) have approximations valid for all time.

In time-independent Hamiltonian systems and in general conservative systems the problem of extension of the timescale of validity is more difficult as these systems do not contain asymptotically stable solutions. For 2 dof Hamiltonian systems we can obtain longtime approximations by using geometric arguments based on the theory of area-preserving maps. See [4] and [18].

We consider another example but with dissipation and forcing.

6.3.4   The Damped Duffing Equation with $O(1)$ Forcing

It is interesting and realistic to modify the Duffing-equation to allow $O(1)$ forcing. Forced oscillations with a nonlinear restoring force present basic problems in nonlinear mechanics leading to much attention. See for instance the book by Stoker [60] ch. 4 with a detailed analysis and emphasis on the use of the harmonic balance method; this choice of analysis makes the results of Stoker mathematical formal, but in [60] the mathematically rigorous Poincaré-Lindstedt method is also partly used.

Consider the Duffing-equation with small viscous damping and $O(1)$ forcing in the form:

$$\ddot{x} + \varepsilon\mu\dot{x} - \varepsilon x^3 + x = h \sin\omega t, \quad \omega \neq 1, h, \mu > 0. \tag{6.14}$$

The restoring force is softening. Following Sect. 1.5 we start with a particular inhomogeneous solution $x_p(t)$ of the reduced equation for $\varepsilon = 0$:

$$x_p(t) = \frac{h}{1 - \omega^2} \sin\omega t, \quad \omega \neq 1. \tag{6.15}$$

With homogeneous solution $r_0 \cos(t + \phi_0)$, this induces us to transform $x, \dot{x} \to r, \phi$ by:

$$\begin{cases} x(t) &= r\cos(t + \phi) + \frac{h}{1-\omega^2}\sin\omega t, \\ \dot{x}(t) &= -r\sin(t + \phi) + \frac{h\omega}{1-\omega^2}\cos\omega t. \end{cases} \tag{6.16}$$

Replacing $-\mu\dot{x} + x^3$ by F we find the variational equations:

$$\dot{r} = -\varepsilon\sin(t + \phi)F, \quad \dot{\phi} = -\frac{\varepsilon}{r}\cos(t + \phi)F, \tag{6.17}$$

with the assumption that a neighbourhood of the origin in polar coordinates $r = 0$ is excluded, and where the arguments of $F(x, \dot{x})$ have to be replaced using transformation (6.16).

In a neighbourhood of $r = 0$ we should use the transformation $x, \dot{x} \to A, B$ by:

$$\begin{cases} x(t) &= A\cos t + B\sin t + \frac{h}{1-\omega^2}\sin\omega t, \\ \dot{x}(t) &= -A\sin t + B\cos t + \frac{h\omega}{1-\omega^2}\cos\omega t, \end{cases} \tag{6.18}$$

leading to the variational system:

$$\dot{A} = -\varepsilon\sin(t)F, \quad \dot{B} = \varepsilon\cos(t)F, \tag{6.19}$$

For this variational system the arguments of $F(x, \dot{x})$ have to be replaced using transformation (6.18).

We use $F = -\mu\dot{x} + x^3$ and consider first a neighbourhood of $x_p(t)$, requiring us to use variational system (6.19). We will use for the averaged quantities again A, B or r, ϕ.

The Case $\omega \neq 1, \frac{1}{3}, 3$. We find after averaging over t:

$$\begin{cases} \dot{A} = -\varepsilon\frac{\mu}{2}A - \varepsilon\frac{3B}{8}(A^2 + B^2 + 2\frac{h^2}{(1-\omega^2)^2}), \\ \dot{B} = -\varepsilon\frac{\mu}{2}B + \varepsilon\frac{3A}{8}(A^2 + B^2 + 2\frac{h^2}{(1-\omega^2)^2}). \end{cases} \qquad (6.20)$$

Multiplying the first equation by A, the second one by B and adding the equations we find:

$$\frac{d}{dt}(A^2 + B^2) = -\varepsilon\mu(A^2 + B^2),$$

so A, B decrease and the phase-flow shrinks exponentially as $\exp(-\varepsilon\mu t)$. if $\omega \neq 1, \frac{1}{3}, 3$ the origin $A = B = 0$ of the averaged system is exponentially stable.

We conclude with Theorem 6.2:

If $\omega \neq 1, \frac{1}{3}, 3$, the solution $x_p(t)$ of (6.15) approximates the solutions of Eq. (6.14) to $O(\varepsilon)$ for all time on a domain containing the origin with size independent of ε.

The Case $\omega = 3$. After averaging over t: using transformation (6.18) we find that $A = B = 0$ is stable in the averaged system. The results are more transparent using (r, ψ) variables with transformation (6.16) for $r > 0$; we checked that the results are equivalent. We find the averaged system:

$$\dot{r} = \varepsilon r(-\frac{1}{2}\mu + \frac{3}{64}hr\cos 3\phi), \quad \dot{\phi} = -\frac{3}{8}\varepsilon(r^2 + \frac{h}{8}r\sin 3\phi + \frac{1}{32}h^2).$$

Critical points for $r > 0$ arise if:

$$\frac{3}{32}hr\cos 3\phi = \mu, \quad r^2 + \frac{h}{8}r\sin 3\phi + \frac{1}{32}h^2 = 0.$$

As the discriminant of the second equation will be negative, we find at this order of approximation no additional subharmonic ($\omega = 3$) periodic solution of size $O(\varepsilon)$ of Eq. (6.14). We conclude with Theorem 6.2:

If $\omega = 3$, the solution $x_p(t)$ of (6.15) approximates the solutions of Eq. (6.14) starting in a neighbourhood of $A = B = 0$ ($r = 0$) to $O(\varepsilon)$ for all time. No subharmonic periodic solutions of system (6.14) exist in this neighbourhood apart from $x_p(t)$.

The Case $\omega = \frac{1}{3}$ As for the case $\omega = 3$ we can average over t using transformation (6.16) or (6.16); again we checked that the results are equivalent. Interestingly the origin is not a solution of the averaged system in this case.

Using (r, ϕ) coordinates we find after averaging:

$$\dot{r} = \frac{\varepsilon}{2}(-\mu r + \frac{729}{2048}h^3\cos\phi), \quad \dot{\phi} = -\frac{3}{8}\frac{\varepsilon}{r}(r^3 + \frac{81}{32}h^2r + \frac{243}{512}h^3\sin\phi), \qquad (6.21)$$

with critical points r_0, ϕ_0. We find for these stationary solutions r_0, ϕ_0:

$$\cos \phi_0 = \frac{2048}{729} \frac{\mu}{h^3} r_0, \quad \sin \phi_0 = -\frac{512}{243} \frac{r_0^3}{h^3} - \frac{16}{3} \frac{r_0}{h}. \tag{6.22}$$

Using the formula $\cos^2 \phi_0 + \sin^2 \phi_0 = 1$ we can derive a cubic equation for $r_0^2 = z$. This equation is of the form:

$$\frac{a}{h^6} z^3 + \frac{b}{h^4} z^2 + \frac{c\mu^2}{h^6} z + \frac{d}{h^2} z = 1, \ a, b, c, d > 0.$$

It is easy to see that for fixed damping coefficient μ and h large enough we will have a unique positive solution $z = r_0^2$ with $|\cos \psi_0| \le 1, |\sin \psi_0| \le 1$. Using expression (6.16) and $x_p(t) = \frac{9}{8} h \sin(t/3)$ we have found an $O(\varepsilon)$ approximation of a periodic solution.

The stability of this periodic solution is obtained by linearising system (6.21) at r_0, ϕ_0. We find the matrix:

$$\begin{pmatrix} -\frac{\varepsilon}{2}\mu & -\frac{1}{2}\varepsilon \frac{729}{2048} h^3 \sin \phi_0 \\ -\frac{3}{8} \frac{\varepsilon}{r_0}(3r_0^2 + \frac{81}{32}h^2) & -\frac{3}{8} \frac{\varepsilon}{r_0} \frac{243}{512} h^3 \cos \phi_0 \end{pmatrix}. \tag{6.23}$$

The eigenvalues are λ_1, λ_2; as $\cos \psi_0 > 0$ the trace of the matrix $(\lambda_1 + \lambda_2)$ is negative. As $\sin \phi_0 < 0$, the product $\lambda_1 \lambda_2$ is positive. We conclude that the critical point and so the periodic solution is asymptotically stable.

Figure 6.1 (left) presents the $x(t)$ time series if $\omega = 1/3$ with initial transient in the case $\mu = h = 1, \varepsilon = 0.1$. Note that we did not study the domains of attraction of the periodic solutions as dependent on μ and h, there may be other types of

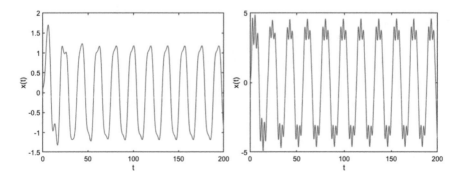

Fig. 6.1 Time series $x(t)$ (200 timesteps) with initial transient in the case of the forced Duffing-equation with damping (6.14) starting at $x(0) = 0.1, \dot{x}(0) = 0.1$ tending to a periodic solution $r_0 \cos(t + \psi_0) + \frac{9}{8} h \sin(t/3)$. We have chosen $\varepsilon = 0.1, \mu = h = 1, \omega = 1/3$. For comparison we present (right figure) the timeseries in the case of hardening force field $+\varepsilon x^3$ with the same data except $h = 10$

solutions. A number of numerical experiments confirm the possibility of unbounded solutions, for instance changing in Fig. 6.1 h from 1 to 2 produces these. On the other hand, with hardening force $+\varepsilon x^3$ we still find periodic solutions for larger forcing amplitudes; in Fig. 6.1 (right) we show the periodic solution in the hardening case with $h = 10$.

Chapter 7
Averaging over Spatial Variables

Consider again the perturbed harmonic Eq. (1.5) or systems of such equations or more general equations describing oscillations that will involve amplitudes and angles. In this context it can also be natural to average over one or more angles instead of over time, but they have to be timelike; for this concept see also Remark 7.2.

7.1 Averaging over One Angle

Theorem 7.1 *Consider the $(n + 1)$-dimensional system:*

$$\dot{x} = \varepsilon X(\phi, x), \ \dot{\phi} = \omega(x) + \varepsilon \ldots \tag{7.1}$$

with $x, x_0 \in D \subset \mathbb{R}^n, 0 \leq \phi \leq 2\pi$ (ϕ 1-dimensional) and X 2π-periodic in ϕ. Introduce averaging over the angle ϕ by:

$$X^0(y) = \frac{1}{2\pi} \int_0^{2\pi} X(s, y)ds.$$

Assume:

1. *The vector functions X, ω and the higher order terms indicated by dots are continuously differentiable on their respective domains of definition.*
2. *The frequency function $\omega(x)$ is bounded away from zero by a constant (ϕ is timelike).*

3. The solutions of the system averaged over the angle

$$\dot{y} = \varepsilon X^0(y), \quad \dot{\phi} = \omega(y), \tag{7.2}$$

remain in an interior subset of D.

then we have $x(t) - y(t) = O(\varepsilon)$ *on the timescale* $1/\varepsilon$.

Remark 7.1 A proof can be found in [58] ch. 7. The proof becomes more complicated if $\omega(x)$ depends explicitly on time. As we shall see, a domain in x-space where the frequency $\omega(x)$ vanishes will be called a *resonance manifold*.

Remark 7.2 An alternative approach to the proof uses that the angle ϕ is timelike. We mean by this that ϕ is a monotonically increasing or decreasing function of time, the time-interval can be mapped one-to-one on the circle that ϕ travels around repeatedly. We can eliminate time in the case of one angle by dividing the equation for x by the equation for ϕ and average over "time" ϕ.

7.2 Averaging over More Angles

The general idea of averaging over more angles is not difficult to understand, but the details of the applications contain many difficulties and surprises. The reader is referred also to an extensive account in ch. 4 of [3].

Consider the system:

$$\dot{x} = \varepsilon X(\phi, x), \quad \dot{\phi} = \Omega(x), \tag{7.3}$$

with $x \in D \subset \mathbb{R}^n$, $\phi \in T^m$. T^m is the m-dimensional torus described by m angles. We assume that the vector field X is periodic in the m angles with in general different periods. Using the individual angles ϕ_i, $i = 2, \ldots, m$ as timelike variables we can average over the individual angles unless a component $\Omega_i(x)$ vanishes in D. However, this picture is not complete, with more than one angle the interaction can be more complicated. To show this consider a generalised Fourier analysis of $X(\phi, x)$ of the form:

$$X(\phi, x) = \sum c_k(x) e^{i(k_1\phi_1 + \ldots + k_m\phi_m)},$$

with index $k = (k_1, \ldots, k_m)$. If for certain k with $c_k(x) \neq 0$ we have that there is a subdomain where the *combination angle* χ_k vanishes:

$$\chi_k(x) = k_1\Omega_1(x) + k_2\Omega_2(x) + \ldots + k_m\Omega_m(x) = 0,$$

then the corresponding combination angle χ_k is not timelike. We will call such a domain a *resonance manifold* for system (7.3). Outside the resonance manifolds in

D Theorem 7.1 for more angles applies, inside and near these manifolds we have to consider system (7.3) separately.

Remark 7.3 (Nonautonomous Equations) In practice, it happens quite often that time t enters explicitly into the equations. Consider the system

$$\dot{x} = \varepsilon X(\phi, t, x),$$
$$\dot{\phi} = \Omega(x),$$

with the vector function X periodic in t. Suppose that ϕ is m-dimensional; we add the angle ϕ_{m+1} with:

$$\phi_{m+1} = t, \quad \dot{\phi}_{m+1} = 1,$$

and consider averaging over $m + 1$ angles.

Example 7.1 To illustrate the concepts we consider a system with 2 angles.

$$\begin{cases} \dot{x} &= \varepsilon(c_0(x) + c_1(x)\cos\phi_1 + c_2(x)\sin 2\phi_2 + c_3(x)\cos(\phi_1 - \phi_2)), \\ \dot{\phi}_1 &= \omega_1(x), \\ \dot{\phi}_2 &= \omega_2(x). \end{cases} \qquad (7.4)$$

The functions $c_0(x), \ldots, c_3(x), \omega_1(x), \omega_2(x)$ are smooth and nontrivial on $\mathbb{R}$. Suppose first that the frequency functions are constant: $\omega_1(x) = k, \omega_2(x) = l$. If $k \neq l$ there is no resonance manifold, the angles ϕ_1, ϕ_2 and the combination angle $(\phi_1 - \phi_2)$ are timelike so we can average to find: $\dot{x} = \varepsilon c_0(x)$.

Suppose now that the frequency functions are not constant but positive, there exists an x_0 such that $\omega_1(x_0) = \omega_2(x_0)$. So if $x(t) = x_0, \dot{\phi}_1 - \dot{\phi}_2 = 0$, we cannot average over $(\phi_1 - \phi_2)$ in a neighbourhood of x_0. A neighbourhood of x_0 is called a resonance manifold M, outside this neighbourhood we find again by averaging $\dot{x} = \varepsilon c_0(x)$. In subsequent examples we will characterise the dynamics inside a resonance manifold. We will also present the case of solutions that are passing though a resonance manifold. In the present example this happens for instance if we put: $c_0(x) = 1, \omega_1(x) = 2, \omega_2(x) = x, x(0) = 1$. System (7.4) becomes:

$$\begin{cases} \dot{x} &= \varepsilon(1 + c_1(x)\cos\phi_1 + c_2(x)\sin 2\phi_2 + c_3(x)\cos(\phi_1 - \phi_2)), \\ \dot{\phi}_1 &= 2, \\ \dot{\phi}_2 &= x. \end{cases} \qquad (7.5)$$

The angle ϕ_1 is timelike, we can average over ϕ_1; near the initial condition $x(0) = 1$ the angle ϕ_2 is also timelike, we can average over ϕ_2 unless at some point x tends to zero. For the combination angle $\psi = \phi_1 - \phi_2$ we have

$$\dot{\psi} = \dot{\phi}_1 - \dot{\phi}_2 = 2 - x.$$

So near $x = 2$ we have a resonance manifold M. Initially we have after averaging over ϕ_1, ϕ_2, ψ the approximation $x(t) = 1 + \varepsilon t$ but we will move into the resonance manifold M at $x = 2$. The questions are then, do we pass M and how, or are the solutions caught in M? A similar question would arise if we start at $x(0) < 0$. Such questions will be raised and answered in the applications. To answer the question in the case of Eq. (7.5) we have to be more explicit about the coefficients $c_1(x), c_2(x), c_3(x)$.

7.3   Applications

In the case of 2-dimensional autonomous systems like the Van der Pol-equation and the Duffing-equation we can easily apply Theorem 7.1; we leave such applications to the reader. It is dynamically interesting to study what happens near resonance manifolds. In the case of more angles this turns out to be quite complicated, we will develop the necessary theory while discussing examples.

7.3.1   A Pendulum with Slow Time Frequency Variation

The transformations we will use in this problem offer a different perspective on the perturbation behaviour. Consider the equation:

$$\ddot{x} + \omega^2(\varepsilon t)x = 0. \tag{7.6}$$

We put $\dot{x} = \omega(\varepsilon t)y$, $\tau = \varepsilon t$; and using a change of transformation (1.21) by computing $\ddot{x} = \dot{\omega}y + \omega\dot{y}$ so that

$$\dot{y} = \frac{\ddot{x}}{\omega} - \frac{\dot{\omega}}{\omega}y = -\omega x - \frac{\dot{\omega}}{\omega}y.$$

we find with $x = r \sin\phi$, $y = r \cos\phi$:

$$\dot{r} = -\varepsilon \frac{1}{\omega(\tau)}\frac{d\omega}{d\tau} r \cos^2\phi,$$

$$\dot{\phi} = \omega(\tau) + \varepsilon \frac{1}{\omega(\tau)}\frac{d\omega}{d\tau} \sin\phi \cos\phi,$$

$$\dot{\tau} = \varepsilon.$$

If we assume that $0 < a < \omega(\tau) < b$ (with a, b constants independent of ε), ϕ is timelike. We have a three-dimensional system periodic in ϕ with two equations slowly varying. The function $\omega(\tau)$ is smooth and its derivative is bounded. Interestingly we have made no other assumptions on $\omega(\varepsilon t)$.

Averaging the 2 slowly varying equations over ϕ, we obtain

$$\dot{r} = -\varepsilon \frac{1}{2\omega(\tau)} \frac{d\omega}{d\tau} r,$$

$$\dot{\tau} = \varepsilon.$$

The equation for τ does not change. From this system, we get

$$\frac{dr}{d\tau} = -\frac{1}{2\omega(\tau)} \frac{d\omega}{d\tau} r,$$

which can be integrated to find

$$r(\tau) = \frac{r(0)\sqrt{\omega(0)}}{\sqrt{\omega(\tau)}}.$$

Note that the quantity $r(\varepsilon t)\sqrt{\omega(\varepsilon t)}$ is conserved in time, it is an *adiabatic invariant* of Eq. (7.6). It is clear that the same technique can be applied to equations of the form

$$\ddot{x} + \varepsilon\mu\dot{x} + \omega^2(\varepsilon t)x + \varepsilon f(x) = 0.$$

7.3.2 A Typical Problem with One Resonance Manifold

Interesting problems may arise when $\omega(x)$ is not bounded away from zero, this will become an important issue in higher-dimensional problems with more angles. To prepare for this, we consider an artificial example.

Consider the two scalar equations

$$\dot{x} = \varepsilon(1 - \cos\phi + \sin 3\phi),$$

$$\dot{\phi} = x - 1.$$

If $x(0) \neq 1$ ϕ is timelike and averaging over ϕ produces

$$\dot{x} = \varepsilon, \ x(0) = x(0),$$

so that $x(t) = x(0) + \varepsilon t$ and $x(t) - (x(0) + \varepsilon t) = O(\varepsilon)$ on the timescale $1/\varepsilon$ as long as $x(0) + \varepsilon t$ remains outside a neighbourhood of $x = 1$. However, if $x(0) < 1$ the approximate solution will pass into a neighbourhood of resonance manifold $x = 1$. Such a neighbourhood is called a *resonance zone*.

As we shall see, a resonance zone is really a boundary layer in the sense that its size tends to zero as ε tends to zero. We can see this when introducing a local variable ξ by rescaling near $x = 1$. We put

$$\xi = \frac{x - 1}{\delta(\varepsilon)},$$

with $\delta(\varepsilon)$ a function of ε to be determined. Transforming $x, \phi \to \xi, \phi$, the system becomes:

$$\dot{\xi} = \frac{\varepsilon}{\delta(\varepsilon)}(1 - \cos\phi + \sin 3\phi),$$

$$\dot{\phi} = \delta(\varepsilon)\xi.$$

the equations are balanced (sometimes called a significant degeneration of the differential operator) if the terms on the righthand side are of the same order in ε or

$$\frac{\varepsilon}{\delta(\varepsilon)} = \delta(\varepsilon).$$

We conclude that $\delta(\varepsilon) = \sqrt{\varepsilon}$, which is the size of the resonance zone. Differentiating the equation for ϕ we can eliminate ξ to find the equation:

$$\ddot{\phi} + \varepsilon(\cos\phi - \sin 3\phi) = \varepsilon.$$

This conservative equation governs the dynamics in the resonance zone with timescale $1/\sqrt{\varepsilon}$.

7.3.3   Behaviour in a Resonance Manifold

The preceding example of a conservative equation is more typical than it seems. Consider for instance the 2-dimensional system consisting of 2 coupled scalar equations:

$$\dot{x} = \varepsilon X(\phi, x),$$

$$\dot{\phi} = \omega(x),$$

with X and ω smooth scalar functions and $X(\phi, x)$ 2π-periodic in ϕ. Averaging over ϕ is possible outside domains determined by the zeros of $\omega(x)$. Suppose $\omega(c) = 0$ with c an isolated zero of $\omega(x)$ and rescale:

$$\xi = \frac{x - c}{\delta(\varepsilon)}.$$

The equations become

$$\delta(\varepsilon)\dot{\xi} = \varepsilon X(\phi, c + \delta(\varepsilon)\xi),$$
$$\dot{\phi} = \omega(c + \delta(\varepsilon)\xi).$$

Expanding we find:

$$\delta(\varepsilon)\dot{\xi} = \varepsilon X(\phi, c) + O(\delta(\varepsilon)\varepsilon),$$
$$\dot{\phi} = \delta(\varepsilon)\frac{d\omega}{dx}(c)\xi + O(\delta^2(\varepsilon)).$$

Again, a suitable balancing (significant degeneration) arises if we choose $\delta(\varepsilon) = \sqrt{\varepsilon}$, which is the size of the resonance zone. The equations in the resonance zone are to first order

$$\dot{\xi} = \sqrt{\varepsilon}X(\phi, c),$$
$$\dot{\phi} = \sqrt{\varepsilon}\frac{d\omega}{dx}(c)\xi,$$

which is again to first order a conservative system (one can show this by differentiating the equation for ϕ and eliminating ξ). We demonstrate this explicitly in a pendulum system.

Example 7.2 Consider the system:

$$\begin{cases} \dot{x} = & \varepsilon(-x - \sin\phi), \\ \dot{\phi} = & x, \end{cases} \tag{7.7}$$

From the preceding analysis we know that a neighbourhood of $x = 0$ is a resonance zone. Outside the resonance zone we average over ϕ to obtain an approximation of $x(t)$ described by:

$$\dot{x} = -\varepsilon x.$$

With given $x(0)$ we find the approximation $x(0)\exp(-\varepsilon t)$. In the resonance zone we rescale $\xi = x/\sqrt{\varepsilon}$ which leads to the system:

$$\sqrt{\varepsilon}\dot{\xi} = \varepsilon(-\sqrt{\varepsilon}\xi - \sin\phi),$$
$$\dot{\phi} = \sqrt{\varepsilon}\xi,$$

We find

$$\dot{\phi} = \sqrt{\varepsilon}\xi, \ \dot{\xi} = -\sqrt{\varepsilon}\sin\phi - \varepsilon\xi,$$

which is to order $O(\sqrt{\varepsilon})$ again conservative, but not to order $O(\varepsilon)$. More insight is obtained by differentiating the equation for ϕ in system (7.7); producing:

$$\ddot{\phi} + \varepsilon\dot{\phi} + \varepsilon \sin\phi = 0.$$

The equation describes the motion of a mathematical pendulum with linear damping. The equilibria at $(2n\pi, 0), n = 1, 2, \ldots$ are positive attractors, but this is described asymptotically by a second order calculation in the resonance zone, not by first order.

7.3.4 A 3-Dimensional System with 2 Angles

As indicated earlier, when more angles are present, many subtle problems and interesting phenomena may arise.

Example 7.3 We start again with an example. Consider the system with given initial value $x(0)$:

$$\dot{x} = \varepsilon x(\cos\phi_1 + \cos(2\phi_1 - \phi_2)),$$
$$\dot{\phi}_1 = x^2 + 1,$$
$$\dot{\phi}_2 = 1.$$

In the equation for x we have angle ϕ_1 and combination angle $2\phi_1 - \phi_2$. Both are timelike as $\dot{\phi}_1 > 0$ and $2\dot{\phi}_1 - \dot{\phi}_2 = 2x^2 + 1 > 0$. Averaging the equation for x over ϕ_1 and the combination angle we find the averaged equation $\dot{x} = 0$. We conclude that averaging over more angles yields that $x(0)$ is an $O(\varepsilon)$ approximation of $x(t)$ on the timescale $O(1/\varepsilon)$.

We consider a modification of the last example.

Example 7.4 Consider the system

$$\begin{cases} \dot{x} &= \varepsilon + \varepsilon x(\cos\phi_1 - 2\cos(2\phi_1 - \phi_2)), \\ \dot{\phi}_1 &= x^2 + 1, \\ \dot{\phi}_2 &= 4. \end{cases} \tag{7.8}$$

We have still the same two angle combinations in the equation for x as in the preceding example. Again $\dot{\phi}_1 > 0$ so ϕ_1 is timelike, but we find $\dot{\psi} = 2\dot{\phi}_1 - \dot{\phi}_2 = 2x^2 - 2$ so the second combination angle is timelike only outside 2 resonance zones which are neighbourhoods of $x = \pm 1$. Outside the resonance zones we have the approximation $x(0) + \varepsilon t$ for $x(t)$, so if $x(0) < 1$ we will enter at least one resonance zone. To describe the dynamics near $x = 1$ we rescale as before:

$$\xi = \frac{x - 1}{\sqrt{\varepsilon}},$$

to find in the resonance zone near $x = 1$:

$$\dot{\xi} = \sqrt{\varepsilon}(1 + \cos\phi_1 - 2\cos\psi) + O(\varepsilon),$$
$$\dot{\phi}_1 = 2 + O(\sqrt{\varepsilon}),$$
$$\dot{\psi} = 4\sqrt{\varepsilon}\xi + O(\varepsilon).$$

Averaging over ϕ_1 and differentiating the equation for the combination angle ψ we find to first order:

$$\ddot{\psi} = 4\varepsilon(1 - 2\cos\psi).$$

The first order equation in the resonance zone is again conservative, but it is interesting to note that we have a centre critical point (equilibrium) ($\psi = -\pi/2$, $\dot{\psi} = 0$) and a saddle ($\psi = \pi/2$, $\dot{\psi} = 0$) in the resonance zone. The saddle will persist as an unstable solution in higher order approximation. The nature of the perturbed centre will show at higher order, see Fig. 7.1.

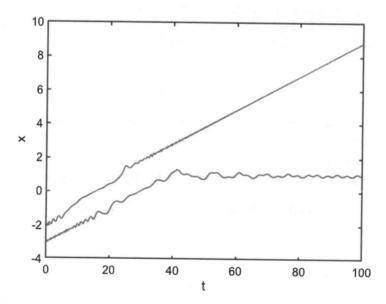

Fig. 7.1 Passing into resonance for system (7.8) with $\varepsilon = 0.1$, initial conditions $x(0) = -2, \phi_1(0) = 1, \phi_2(0) = 0$ and $x(0) = -3, \phi_1(0) = 1, \phi_2(0) = 0$. In the second case the attraction is into the resonance zone $x = 1$

In some of the preceding examples we encountered the phenomenon of solutions that are attracted to an equilibrium in a resonance zone. We call this *resonance locking*.

7.3.5 Intersection of Resonance Manifolds

We analyse a more complicated example with two slowly varying components and two angles where we have resonance domains that may intersect.

Consider the system:

$$\begin{cases} \dot{x}_1 &= \varepsilon(0.5 + 2x_1 \cos \phi_1 + x_2 \cos \phi_2), \\ \dot{x}_2 &= \varepsilon(0.5 - x_1 \cos \phi_1 + 1.5x_2 \cos \phi_2), \\ \dot{\phi}_1 &= x_2 - x_1^2, \\ \dot{\phi}_2 &= x_1 - 1. \end{cases} \tag{7.9}$$

The angle ϕ_1 is timelike outside the set of points where $x_2 = x_1^2$; ϕ_2 is timelike outside the set of points where $x_1 = 1$. Averaging outside the two resonance manifolds produces the approximations $x_1(t) = x_1(0) + \varepsilon t$ and $x_2(t) = x_2(0) + \varepsilon t$. This follows from averaging theory and it is for initial values outside the resonance zones confirmed by numerical computation.

A double resonance arises at the intersection $x_1 = x_2 = 1$.

The analysis at the separate resonance manifolds runs as before. Leaving out the location of the double resonance we put respectively:

$$\frac{x_1 - 1}{\sqrt{\varepsilon}} = \xi, \text{ and } \frac{x_2 - x_1^2}{\sqrt{\varepsilon}} = \eta$$

to perform the rescalings of the equations at the two sets. We discuss the dynamics in the separate manifolds.

The Resonance Manifold $x_1 = 1$
Assume $x_2 \neq 1$. The equations for x_1, x_2 can be averaged over ϕ_1 producing:

$$\dot{x}_1 = \varepsilon(0.5 + x_2 \cos \phi_2), \ \dot{x}_2 = \varepsilon(0.5 + 1.5x_2 \cos \phi_2),$$

with integral

$$\frac{3}{2}x_1(t) - x_2(t) = \frac{\varepsilon}{4}t + \frac{3}{2}x_1(0) - x_2(0).$$

If $x_1(t)$ remains in the resonance set we should have

$$x_2(t) = -\frac{\varepsilon}{4}t + x_2(0) + \ldots$$

where the dots stand for higher order terms. This is not compatible with the equation for x_2 so we conclude that $x_1(t)$ leaves the resonance manifold. This is confirmed by a number of numerical experiments.

The Resonance Manifold $x_2 = x_1^2$

As we leave out a neighbourhood of $(x_1, x_2) = (1, 1)$, averaging over ϕ_2 yields that the terms with ϕ_2 in system (7.9) vanish at first order. Choosing $x_1 = x_r \neq 1, x_2 = x_r^2$, we obtain after rescaling:

$$\dot{\eta} = \sqrt{\varepsilon}(\frac{1}{2}(1 - 2x_r) - (x_r + 4x_r^2)\cos\phi_1) + O(\varepsilon), \dot{\phi}_1 = \sqrt{\varepsilon}\eta.$$

Differentiating the equation for ϕ_1 we get:

$$\ddot{\phi}_1 + \varepsilon(x_r + 4x_r^2)\cos\phi_1 = \varepsilon\frac{1}{2}(1 - 2x_r) + O(\varepsilon^{\frac{3}{2}}). \qquad (7.10)$$

If $x_r < -\frac{1}{4} - \frac{1}{4}\sqrt{3}$ or $x_r > -\frac{1}{4} + \frac{1}{4}\sqrt{3}, x_r \neq 1$, Eq. (7.10) has a centre and a saddle equilibrium. See Fig. 7.2 where the behaviour of $\eta(t)$ and $x_2 - x_1^2$ is presented. The first order equation (7.10) becomes with the initial condition $x_1(0) = 0.5, x_2(0) = 0.25$:

$$\ddot{\phi}_1 + \frac{3}{2}\varepsilon\cos\phi_1 = 0, \text{ critical points } \phi_1 = \pm\pi/2.$$

It is remarkable that with the initial conditions of Fig. 7.2 in the resonance zone, the solutions follow the resonance zone to end up in the double resonance zone. Keeping the initial values $x_1(0) = 0.5, x_2(0) = 0.25, \phi_1(0) = -1.57$, the same pattern of ending up in the double resonance zone is obtained when choosing $\phi_2(0)$ in the interval I given by $-0.05 \leq \phi_2(0) \leq 0.88$.

Solutions that are starting close to the trapping interval I are interacting for some time in the double resonance zone followed by escape from this zone; see Fig. 7.3.

The Double Resonance

Near the double resonance we use the rescaling (blowup):

$$\frac{x_1 - 1}{\sqrt{\varepsilon}} = \xi, \frac{x_2 - 1}{\sqrt{\varepsilon}} = \eta,$$

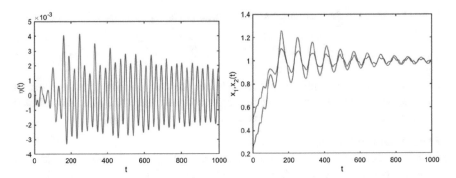

Fig. 7.2 Dynamics starting in the resonance set $x_2 = x_1^2$. Time series $\eta(t)$ (1000 timesteps) for system (7.9) starting left at $x_1(0) = 0.5$, $x_2(0) = 0.25$, $\phi_1(0) = -1.57$, $\phi_2(0) = 0.5$ with $\varepsilon = 0.01$. The solutions are starting near an unstable solution and tend to a stable solution in the double resonance set near $x_1 = x_2 = 1$ (right picture)

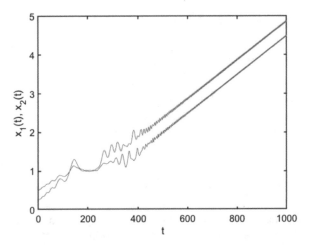

Fig. 7.3 Dynamics starting in the resonance manifold $x_2 = x_1^2$. Time series (1000 timesteps) for $x_1(t), x_2(t)$ of system (7.9) starting at $x_1(0) = 0.5$, $x_2(0) = 0.25$, $\phi_1(0) = -1.57$, $\phi_2(0) = -0.06$ with $\varepsilon = 0.01$. The solutions are close to being trapped in the double resonance zone near $x_1 = x_2 = 1$ but are leaving this zone after about 200 timesteps

producing with system (7.9):

$$
\begin{cases}
\dot{\xi} & = \sqrt{\varepsilon}(0.5 + 2\cos\phi_1 + \cos\phi_2) + \varepsilon \ldots, \\
\dot{\eta} & = \sqrt{\varepsilon}(0.5 - \cos\phi_1 + 1.5\cos\phi_2) + \varepsilon \ldots, \\
\dot{\phi}_1 & = \sqrt{\varepsilon}(\eta - 2\xi) + \varepsilon\xi^2, \\
\dot{\phi}_2 & = \sqrt{\varepsilon}\xi.
\end{cases}
\tag{7.11}
$$

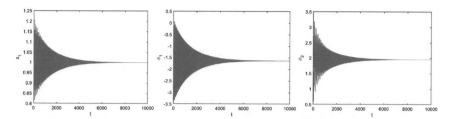

Fig. 7.4 Time series (10,000 timesteps) for system (7.9) starting at $x_1(0) = 1, x_2(0) = 0.95, \phi_1(0) = 0.1, \phi_2(0) = 1.6$ with $\varepsilon = 0.01$. These solutions tend to a stable solution in the double resonance zone near $x_1 = x_2 = 1$.

Differentiation of the equations for ϕ_1, ϕ_2 yields:

$$\ddot{\phi}_1 = -\varepsilon(0.5 + 5\cos\phi_1 + 0.5\cos\phi_2) + O(\varepsilon^{3/2}),$$
$$\ddot{\phi}_2 = \varepsilon(0.5 + 2\cos\phi_1 + \cos\phi_2) + O(\varepsilon^{3/2}). \tag{7.12}$$

The first order equations generate a volume-preserving flow in phase-space; in Fig. 7.4 a numerical calculation of system (7.9) shows convergence to a stable solution in the double resonance zone.

An independent check is the analysis of critical points (equilibria) of autonomous system (7.9). The 4 critical points are in the double resonance zone and are determined by:

$$(x_1, x_2) = (1, 1), \ \cos\phi_1 = -\frac{1}{16}, \cos\phi_2 = -\frac{3}{8}. \tag{7.13}$$

We abbreviate $S_1 = \sin\phi_1$, $S_2 = \sin\phi_2$. The matrix determining the stability at the 4 critical points of system (7.9) is:

$$\begin{pmatrix} -\frac{1}{8}\varepsilon & -\frac{3}{8}\varepsilon & -2\varepsilon S_1 & -\varepsilon S_2 \\ \frac{1}{16}\varepsilon & -\frac{3}{16}\varepsilon & \varepsilon S_1 & -\frac{3}{2}\varepsilon S_2 \\ -2 & 1 & 0 & 0 \\ 1 & 0 & 0 & 0 \end{pmatrix}.$$

We find one equilibrium with negative real eigenvalues and so stability if $\sin\phi_1 < 0$, $\sin\phi_2 > 0$; the eigenvalues depend on ε, we find for instance:

$$\varepsilon = 0.1; \text{ eigenvalues} - 0.0065 \pm 0.7217i \ - 0.0279 \pm 0.2651i,$$
$$\varepsilon = 0.01; \text{ eigenvalues} - 0.0007 \pm 0.2282i \ - 0.0028 \pm 0.0843i ,$$
$$\varepsilon = 0.001; \text{ eigenvalues} - 0.0001 \pm 0.0722i \ - 0.0003 \pm 0.0267i.$$

This settles the question of positive attraction in the double resonance zone.

7.3.6   A Rotating Flywheel on an Elastic Foundation

We have seen the phenomenon of locking into resonance. We consider now an interesting application involving passage through resonance and the possibility of (undesirable) capture into resonance. Consider a spring, modeling an elastic foundation, that can move in the vertical x direction on which a rotating wheel is mounted; the rotation angle is ϕ. The wheel has a small mass fixed on the edge that makes it slightly eccentric, a flywheel; see Fig. 7.5. The vertical displacement x and the rotation ϕ are determined by the equations:

$$\ddot{x} + x = \varepsilon(-x^3 - \dot{x} + \dot{\phi}^2 \cos \phi) + O(\varepsilon^2),$$

$$\ddot{\phi} = \varepsilon \left(\frac{1}{4}(2 - \dot{\phi}) + (1 - x) \sin \phi \right) + O(\varepsilon^2).$$

See [22] for the equations; we have added an appropriate scaling, assuming that the friction, the nonlinear restoring force, the eccentric mass, and several other forces are small. To obtain a standard form suitable for averaging, we transform:

$$x = r \sin \phi_2, \ \dot{x} = r \cos \phi_2, \ \phi = \phi_1, \ \dot{\phi}_1 = \Omega,$$

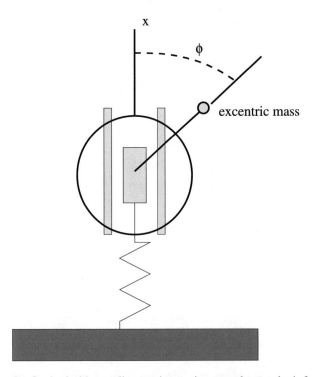

Fig. 7.5   A rotating flywheel with a small eccentric mass is mounted on an elastic foundation

with $r > 0$, $\Omega > 0$. This introduces two angles and two slowly varying quantities:

$$\dot{r} = \varepsilon \cos \phi_2 (-r^3 \sin^3 \phi_2 - r \cos \phi_2 + \Omega^2 \cos \phi_1),$$

$$\dot{\Omega} = \varepsilon \left(\frac{1}{4}(2 - \Omega) + \sin \phi_1 - r \sin \phi_1 \sin \phi_2 \right),$$

$$\dot{\phi}_1 = \Omega,$$

$$\dot{\phi}_2 = 1 + \varepsilon \left(r^2 \sin^4 \phi_2 + \frac{1}{2} \sin 2\phi_2 - \frac{\Omega^2}{r} \cos \phi_1 \sin \phi_2 \right).$$

The $O(\varepsilon^2)$ terms have been omitted. In the equation for r and Ω to $O(\varepsilon)$, the angles are $\phi_1, \phi_2, \phi_1 + \phi_2, \phi_1 - \phi_2$. As ϕ_1 and ϕ_2 are monotonically increasing and so timelike, the only resonance zone that can arise is at this order $\phi_1 - \phi_2 = 0$, which determines the resonance manifold $\Omega = 1$. Outside the resonance zone, a neighbourhood of $\Omega = 1$, we average over the angles to find the approximations given by:

$$\dot{r} = -\frac{1}{2}\varepsilon r,$$

$$\dot{\Omega} = \frac{1}{4}\varepsilon(2 - \Omega).$$

This is already an interesting result. Outside the resonance zone, $r(t)$ will decrease exponentially with time; on the other hand, $\Omega(t)$ will tend to the value 2. If we start with $\Omega(0) < 1$, $\Omega(t)$ will after some time enter the resonance zone around $\Omega = 1$. How does this affect the dynamics? Will the system pass in some way through resonance or will it stay in the resonance zone, resulting in vertical oscillations that are undesirable for a mounted flywheel. Such phenomena are illustrated in Fig. 7.6 where capture in the resonance zone $\Omega = 1$ and transition is shown.

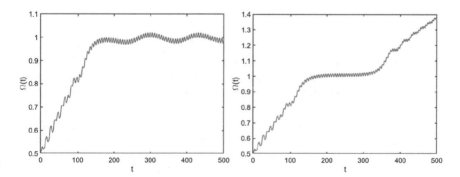

Fig. 7.6 Dynamics of the flywheel on an elastic foundation. Left capture in resonance near $\Omega = 1$ with initial conditions $r(0) = 1$, $\Omega(0) = 0.5$, $\phi_1(0) = 1$, $\phi_2(0) = 0$ and $\varepsilon = 0.01$. With initial conditions $r(0) = 1$, $\Omega(0) = 0.5$, $\phi_1(0) = 0.9600$, $\phi_2(0) = 0$ we have a slow transition through the resonance zone $\Omega = 1$ (right picture)

The way to answer these questions systematically is to analyse what is going on in the resonance zone and find out whether there are attractors present. Following the analysis of localising into the resonance zone as before, we introduce the resonant combination angle $\psi = \phi_1 - \phi_2$ and the local variable

$$\omega = \frac{\Omega - 1}{\sqrt{\varepsilon}}.$$

Transforming the equations for r, Ω and the angles, the leading terms are $O(\sqrt{\varepsilon})$; we find

$$\dot{r} = \varepsilon \cdots,$$
$$\dot{\omega} = \sqrt{\varepsilon}\left(\frac{1}{4} + \sin\phi_1 - \frac{1}{2}r\cos\psi + \frac{1}{2}r\cos(2\phi_1 - \psi)\right) + \varepsilon \cdots,$$
$$\dot{\psi} = \sqrt{\varepsilon}\omega + \varepsilon \cdots,$$
$$\dot{\phi}_1 = 1 + \sqrt{\varepsilon}\omega.$$

We can average over the remaining angle ϕ_1; as the equation for r starts with $O(\varepsilon)$ terms, we have in the resonance zone that $r(t) = r(0) + O(\sqrt{\varepsilon})$. The equations for the first order approximations of ω and ψ are

$$\dot{\omega} = \sqrt{\varepsilon}\left(\frac{1}{4} - \frac{1}{2}r\cos\psi\right),$$
$$\dot{\psi} = \sqrt{\varepsilon}\omega.$$

By differentiation of the equation for ψ, we can write this as the pendulum equation

$$\ddot{\psi} + \frac{1}{2}\varepsilon r(0)\cos\psi = \frac{1}{4}\varepsilon.$$

The timescale of the dynamics is clearly $\sqrt{\varepsilon}t$; there are two equilibria, one a centre point and the other a saddle. They correspond with periodic solutions in the resonance zone of the original system. The saddle is definitely unstable, and for the centre point we have to perform higher-order averaging, to $O(\varepsilon)$, to determine the stability. This analysis was carried out by Van den Broek [67]; see also [66]. The result is that by a second order calculation in the small parameter, the centre point in the resonance zone becomes an attracting focus so that the corresponding periodic solution is stable.

The next (difficult) step is to determine the set of initial values that lead to capture. In [67] an approach, formulated by Haberman in [32], was used as follows. Compute a second order approximation of the solutions outside the resonance zone and determine the values assumed by the solutions on entering the resonance zone. Using the second order approximation in the resonance zone we find that the invariant manifolds of the saddle will have different 'energy' values; these will be used to localise initial values from entering the resonance zone that lead to transition of the zone and capture into resonance.

The implication is that for certain initial values the oscillator-flywheel might pass into resonance and stays there. Van den Broek [67] identified three sets of initial values leading to capture into resonance. The asymptotic estimates are close to the numerical results; in the case $\varepsilon = 0.01$ and with fixed $r(0) = 1, \Omega(0) = 0.5, \phi_2(0) = 0$ we find numerically the 3 sets for ϕ_1 values leading to capture:

$$[0.962, 1.227] \quad [2.845, 3.148] \quad [4.764, 4.923].$$

It is surprising that there are 3 sets; it raises the question whether higher order approximations might produce even more sets, presumably of smaller size.

Chapter 8
Hamiltonian Resonances

This chapter demonstrates the use of averaging for Hamiltonian systems in some important cases. General introductions to Hamiltonian dynamics can be found in [3, 4, 18] and [84]. In this book we emphasise quantitative aspects but of course qualitative aspects cannot be ignored to obtain general insight. In dynamics, in particular Hamiltonian dynamics, symmetries play an important part; a nice introduction is ch. 10 in [28].

The analysis and approximation of Hamiltonian dynamics is often carried out by computing *normal forms*. Normalisation of Hamiltonian vector fields is closely related to averaging, it is often preferred as the process can be formulated to conserve the Hamiltonian character at each approximation step. However, also using averaging the conservative character of the dynamics can be saved so there need no qualitative difference arise between normal forms and averaging results; see [58] for a detailed comparison. In any case, averaging has the advantage of producing validity of approximations on certain timescales.

In Hamiltonian systems many different timescales may play a part. In application 2.3.8 we considered a Hamiltonian 2 dof system in 1:2 resonance, the dominant timescale is $1/\varepsilon$, the timelike variables $t, \varepsilon t$. In application 6.8 we have the 1:1 Hamiltonian resonance and timescale $1/\varepsilon^2$, timelike variables $t, \varepsilon^2 t$. Chapter 7 showed us resonances with timelike variables $\varepsilon^q t$ where q is a rational number. We will meet these again for higher order resonance.

We will discuss various aspects of Hamiltonian examples (see also [58] ch. 10) with attention to different variational formulations.

It is important to keep in mind that in Hamiltonian systems we have no attraction but recurrent motion if the energy manifold is bounded. Roughly stated this means that for a Hamiltonian system on a bounded domain nearly all orbits return an infinite number of times arbitrarily close to the initial conditions. Passage of resonance zones will be characterised by complicated dynamics involving motion around tori and passage near saddle points but not by capture of orbits. In such complicated dynamics one expects delay of recurrence, see [76], and [82].

© The Author(s), under exclusive license to Springer Nature Switzerland AG 2023   107
F. Verhulst, *A Toolbox of Averaging Theorems*, Surveys and Tutorials in the Applied
Mathematical Sciences 12, https://doi.org/10.1007/978-3-031-34515-9_8

The recurrence Theorem 8.2 is important but has not been used much in practical problems. The reason is probably that many interacting particles and complex dynamics may extend the recurrence time enormously. In this chapter we will discuss higher order resonance and show the use of the recurrence theorem in a few 3 dof problems.

8.1   Frequencies and Resonances in the Hamiltonian Case

Consider time-independent Hamiltonian systems H near stable equilibrium. With n dof (degrees-of-freedom) and basic frequencies $\omega_i, i = 1, \ldots, n$ we write:

$$H = \sum_{i=1}^{n} \frac{1}{2}(p_i^2 + \omega_i^2 x_i^2) + H_3(p_1, \ldots p_n, x_1, \ldots, x_n)$$

$$+ H_4(p_1, \ldots p_n, x_1, \ldots, x_n) + \ldots, \tag{8.1}$$

with H_3 a cubic homogeneous polynomial in its arguments, H_4 a quartic homogeneous polynomial, etc. The (first) sum of quadratic terms is called H_2. If $H_3, H_4, \ldots$ depend on x only, we call the Hamiltonian "derived from a potential". Note that instead of variables $p_i, x_i, i = 1, \ldots, n$ standard use is p for the momenta and q for the positions. In the literature all these notations are mixed and we have to follow this slightly confusing custom.

An equivalent form for H_2 is:

$$H_2 = \sum_{i=1}^{n} \frac{1}{2}\omega_i(p_i^2 + x_i^2).$$

If one wants to study the dynamics near equilibria with mixed positive and negative real parts of the eigenvalues (unstable equilibria) or cases with negative and positive ω_i, the last form is preferred. In H_2 the masses m_i are not dependent on ε and were scaled to 1. One can introduce the masses explicitly by rescaling p_i.

The equations of motion induced by Hamiltonian (8.1) are:

$$\dot{x}_i = \frac{\partial H}{\partial p_i}, \quad \dot{p}_i = -\frac{\partial H}{\partial x_i}, i = 1, \ldots, n.$$

If we have a potential problem for Hamiltonian (8.1), the equations of motion correspond with coupled harmonic equations of the form

$$\ddot{x}_i + \omega_i^2 x_i = \ldots$$

with the dots standing for quadratic, cubic etc. polynomial terms in $x_i, i = 1, \ldots, n$.

The Order of Resonance

As we will show explicitly in the examples, to be 'near equilibrium at the origin' is expressed by rescaling $x_i \to \varepsilon x_i, i = 1, \ldots, n$ followed by dividing the equations by ε. We will formulate variational equations. In the case of coupled harmonic equations, the polynomial righthand sides produce Fourier expansions, resonance zones with periodic solutions and possibly invariant manifolds will emerge for rational relations of the frequencies ω_i. We summarise a few results from [58] ch. 10 for prominent resonances. For the frequencies $\omega_1, \ldots, \omega_n$ we have a resonance if for some $k_1, k_2, \ldots, k_n$ with $k_i \in \mathbb{Z}$ we have:

$$k_i \omega_1 + k_2 \omega_2 + \ldots + k_n \omega_n = 0. \tag{8.2}$$

It makes sense to choose the numbers k_i relative prime.

2 dof: low order resonance at H_3 if $\omega_1 : \omega_2 = 1{:}2$; resonance at H_4 if $\omega_1 : \omega_2 = 1{:}1$ or $1{:}3$.

2 dof: higher order resonance if $\omega_1 : \omega_2 = k_1 : k_2, k_1 + k_2 > 4$ (see next section).

3 dof: 4 resonances at H_3 if $(\omega_1 : \omega_2 : \omega_3) = (1{:}2{:}1), (1{:}2{:}2), (1{:}2{:}3), (1{:}2{:}4)$.

3 dof 12 resonances at $H_3 + H_4$ if $(\omega_1 : \omega_2 : \omega_3) = (1{:}1{:}1), (1{:}1{:}3), (1{:}2{:}5), (1{:}2{:}6), (1{:}3{:}3), (1{:}3{:}4), (1{:}3{:}5), (1{:}3{:}6), (1{:}3{:}7), (1{:}3{:}9), (2{:}3{:}4), (2{:}3{:}6)$.

Applications 2.3.8 and 6.8 contain examples of low order 2 dof resonances. They can be handled by first and second order averaging. Note that symmetries in the equations of motion may change the part played by resonances drastically. We have seen this in Ch. 5, Sect. 5.3.5 for evolution to symmetry.

If the frequencies are constant (as assumed in this chapter), it will become clear that the use of angles instead of phases makes little difference. The combination angles that arise in the resonance zones can in this case be replaced by combination phases.

In the applications we will also discuss a few embedded double resonances in 3 dof, the (2:2:3) and (1:1:4) resonances; they are a mix of low and higher order resonances.

8.2   Higher Order Resonance in 2 Degrees-of-Freedom

Consider systems with 2 degrees-of-freedom (dof), frequencies ω_1, ω_2 and

$$k\omega_1 + l\omega_2 = 0.,$$

with k and l relative prime. The case $|k| + |l| \geq 5$ is called a higher order resonance, the corresponding higher order theory was developed mainly for Hamiltonian systems but the ideas apply also to dissipative systems. The theory of *higher order resonance* for 2 dof Hamiltonian systems was started in [56, 64]; see also [58] ch. 10.6.4, for an introduction [57].

We consider 2 dof Hamiltonian systems with basic $k : l$ higher order resonance. It happens in some cases because of symmetries that a lower order resonance is also described by the theory of higher order resonance as symmetries may produce certain degenerations in the averaged system, see [64].

We will see that in the case of higher order resonance the essential perturbation terms are appearing at much higher order in the expansion of the Hamiltonian. We expand again the Hamiltonian $H(p, q)$ in homogeneous polynomials near stable equilibrium at the origin of phase-space with rescaled coordinates:

$$H(p, q) = \frac{1}{2}\omega_1(p_1^2 + q_1^2) + \frac{1}{2}\omega_2(p_2^2 + q_2^2) + \varepsilon H_3 + \varepsilon^2 H_4 + \dots \qquad (8.3)$$

The averaging process, in this context often called *averaging-normalisation*, removes the cubic terms to produce the normal form Hamiltonian $\bar{H}$. As in [58] ch. 10 we will often use the canonical action-angle coordinates

$$\tau = \frac{1}{2}(p^2 + q^2)$$

and ϕ. It turns out that at higher order resonance the quartic terms, quadratic in τ, do not depend on the angles. They arise at much higher order after averaging:

$$\begin{cases} \bar{H} & = \omega_1\tau_1 + \omega_2\tau_2 + \varepsilon^2(A\tau_1^2 + 2B\tau_1\tau_2 + C\tau_2^2) + \dots + \\ & \varepsilon^{k+l-2}|D|\sqrt{2\tau_1}(2\tau_2)^{\frac{k}{2}}\cos(l\phi_1 - k\phi_2 + \alpha) + O(\varepsilon)^{k+l-1}. \end{cases} \qquad (8.4)$$

with constants $|D|, \alpha, A, B, C$; the dots represent terms of size smaller than $O(\varepsilon^2)$ but not containing the combination angle $\chi = l\phi_1 - k\phi_2 + \alpha$. The first resonant term arrives from H_{k+l} at $O(\varepsilon^{k+l-2})$, and depends on the actions τ_1, τ_2 and angle χ. In the analysis both averaging over time t and combination angle χ plays a part. For the combination angle χ we have:

$$\dot{\chi} = 2\varepsilon^2 \left(l(A\tau_1 + B\tau_2) - k(B\tau_1 + C\tau_2)\right) + \dots + \varepsilon^{k+l-2}g(\tau_1, \tau_2)\sin\chi, \qquad (8.5)$$

with g a function of the actions.

Analysing the averaged equations it turns out there are two domains in phase-space where the dynamics is very different, they are characterised by different timelike variables.

Theorem 8.1 *Assume $\omega_1, \omega_2 > 0$ and smoothness of the Hamiltonian to an order as high as necessary. From the theory of averaging over angles we identify a resonance manifold M that is generically determined by the condition*

$$l(A\tau_1 + B\tau_2) - k(B\tau_1 + C\tau_2) = 0.$$

If this equation has a solution on a bounded energy manifold it determines a resonance manifold M. We have:

- *The* resonance domain *(or zone)* D_I *is a neighbourhood of the resonance manifold M. In terms of singular perturbations, the neighbouring resonance zone is the* inner boundary layer. *Introducing the distance $d(P, M)$ for a point P in the resonance zone to the manifold M we have:*

$$D_I = P\{|d(P, M) = O(\varepsilon^{\frac{k+l-4}{2}})\}, \quad k + l \geq 5.$$

- *The remaining part of phase-space, outside the resonance zone, is D_o, the* outer domain. *In the domain D_o, there is, to a certain approximation, hardly any exchange of energy between the two degrees of freedom.*
- *From Hamiltonian (8.5) we find that variations of χ take place at $O(\varepsilon)^{k+l-2}$ with factor $\sin \chi$. Periodic solutions, if present, can be found in the resonance manifold.*

Remark 1 To compute the averaged system to such a high order is quite an effort. However, one can much easier recover some quantitative aspects by a result formulated in [64]: *If the Hamiltonian* (8.3) *is derived from a potential function then we have $\alpha = 0$ for the constant in the combination angle χ of averaged (normalised)* Hamiltonian (8.4).

Remark 2 Higher order resonance in 3 or more dof is largely unexplored. We will discuss a few applications where first order and higher order resonance arises in 3 dof Hamiltonian systems. We will meet in this context the presence of embedded resonance manifolds.

8.3   The Poincaré Recurrence Theorem

This section is based on Poincaré's original text [50] and on [80]. It is included as it presents a tool to characterise conservative dynamical systems of fairly high dimensions.

The phase-flow induced by a time-independent Hamiltonian is volume-preserving, see [76]. The recurrence theorem implies, loosely formulated, that for Hamiltonian systems on a bounded energy manifold, nearly all solutions return after a finite time T_r arbitrarily close to their original position in phase- space.

The recurrence theorem applies quite generally to systems with measure-preserving flow but we restrict ourselves to an important case, time-independent Hamiltonian systems. We have:

Theorem 8.2 *The Poincaré Recurrence Theorem.*

Consider the phase-flow F of a dynamical system induced by the Hamiltonian function $H(p, q)$ with n dof (2n equations of motion). A bounded energy manifold

*M determined by $H(p, q) = E_0$ (positive constant) has a subset $D_0 \subset M$
with positive measure. Consider solutions of Hamiltonian (8.1) with solution
$(p(t), x((t))$. Nearly all initial points $P = (p(0), x((0))$ in D_0 (in the sense of
measure theory) will return under the map F after a finite time arbitrarily close to
initial P in D_0.*

Remark 3 A set consisting of points that do not return near the initial points under
the map F has measure zero. Examples of initial points that do not return are starting
points of homoclinic and heteroclinic solutions. Other exceptional solutions with
this property will be called *wandering*.

For a 1 dof system on a bounded domain, recurrence is trivial in the phase-plane.
For 2 dof systems that are integrable, recurrence behaviour is relatively simple near
a stable periodic solution, but in general this is already not so easy for chaotic 2 dof
systems. Consider the Euclidean norm; for the distance $d(t)$ of solution $(p(t), x(t))$
to the initial point $(p(0), x(0))$ in phase-space we put:

$$d(t) = \sqrt{\sum_{i=1}^{n} \left((p_i(t) - p_i(0))^2 + (x_i(t) - x_i(0))^2\right)}. \tag{8.6}$$

A recurrence time T_r arises if we choose a fixed distance $d_0 > 0$ with $d(T_r) \leq
d_0$. The Euclidean distance $d(t)$ does not produce a fixed time of recurrences
for a fixed value of d_0, we will find a number of times for which the domain
around $(p(0), x(0))$ with size d_0 is traversed. The recurrence time T_r for a given
Hamiltonian system is not a fixed number, it depends on its specific phase-flow and
the chosen initial position in phase-space.

In [80] the following (rough) upper bound is derived for the recurrence time T_r:

Theorem 8.3 *Suppose that we consider an energy manifold of system (8.1) that
is contained in a sphere with radius $R = \sqrt{E_0}$, d_0 is the chosen distance from
$(p(0), x(0))$ in phase-space. Then we have for the recurrence time*

$$T_r \leq \left(\frac{E_0^{n-1/2}}{d_0^{2n-1}}\right) \text{ as } d_0 \to 0.$$

Suppose that for a problem with 3 dof we have $E_0 = 1, d_0 = 0.1$ (fairly
large), then we have already $T_r \leq 10^5$ timesteps. Considering the timeseries (8.6)
numerically will give an indication of the complexity of the phase-flow.

8.4   Applications

Most applications of higher order resonance arose originally from celestial mechanics and galactic dynamics but the equations formulated here have general signif-

icance. As an introduction we consider first an example that is presented as a low order resonance; but to treat it like low order depends on the coefficients. It illustrates the importance of symmetries.

8.4.1 A General Cubic Potential

Consider the 2 dof potential problem with Hamiltonian:

$$H = \frac{1}{2}(\dot{x}^2 + 4x^2 + \dot{y}^2 + y^2) - (\frac{1}{3}a_1 x^3 + a_2 xy^2 + \frac{1}{3}a_3 y^3 + a_4 x^2 y), \qquad (8.7)$$

with a_1, a_2, a_3, a_4 constants, the system is in 2:1 resonance. In a neighbourhood of the origin we scale $x = \varepsilon\bar{x}$ etc., divide by ε and leave out the bars. This produces the equations of motion:

$$\begin{cases} \ddot{x} + 4x &= \varepsilon(a_1 x^2 + a_2 y^2 + 2a_4 xy), \\ \ddot{y} + y &= \varepsilon(2a_2 xy + a_3 y^2 + a_4 x^2). \end{cases} \qquad (8.8)$$

The system was discussed by several authors, for an amplitude-phase treatment if $a_2 \neq 0$ see [74]; we summarise these results. After this summary we will repeat the analysis partly with amplitude-angle transformation. Using transformation (1.6) in the form:

$$x = r_1 \cos(2t+\psi_1), \dot{x} = -2r_1 \sin(2t+\psi_1), y = r_2 \cos(t+\psi_2), \dot{y} = -r_2 \sin(t+\psi_2).$$

we obtain from [74] with $\chi = \psi_1 - 2\psi_2$ the first order averaged system:

$$\begin{cases} \dot{r}_1 &= -\varepsilon\frac{a_2}{8}r_2^2 \sin\chi, \dot{\psi}_1 = -\varepsilon\frac{a_2}{8}\frac{r_2^2}{r_1} \cos\chi, \\ \dot{r}_2 &= \varepsilon\frac{a_2}{2}r_1 r_2 \sin\chi, \dot{\psi}_1 = -\varepsilon\frac{a_2}{2}r_1 \cos\chi, \end{cases} \qquad (8.9)$$

It is remarkable that the coefficients a_1, a_3, a_4 play no part in an $O(\varepsilon)$ approximation on the timescale $1/\varepsilon$. The system (8.9) has 2 independent integrals of motion and 3 periodic solutions for each value of the energy: the x-normal mode corresponding with $y = \dot{y} = 0$(unstable), the in-phase solution for $\chi = 0$ (stable) and the out-phase solution for $\chi = \pi$ (stable). As the energy is a free parameter, we have 3 families of periodic solutions. Interestingly, the y-normal mode can not be continued.

We take now a different approach of [74] to show both the similarity but also slight difference with averaging over angles. Using the amplitude-angle transformation (1.21):

$$x = r_1 \sin\phi_1, \dot{x} = 2r_1 \cos\phi_1, y = r_2 \sin\phi_2, \dot{y} = r_2 \cos\phi_2,$$

we obtain the variational system:

$$\begin{cases} \dot{r}_1 = \frac{\varepsilon}{2}\cos\phi_1(a_1 r_1^2 \sin^2\phi_1 + a_2 r_2^2 \sin^2\phi_2 + 2a_4 r_1 r_2 \sin\phi_1 \sin\phi_2),\ \dot{\phi}_1 = 2 + \varepsilon\ldots, \\ \dot{r}_2 = \varepsilon\cos\phi_2(2a_2 r_1 r_2 \sin\phi_1 \sin\phi_2 + a_3 r_2^2 \sin^2\phi_2 + a_4 r_1^2 \sin^2\phi_1),\ \dot{\phi}_2 = 1 + \varepsilon\ldots. \end{cases}$$

$$(8.10)$$

It is easy to see that the angles in the system for (r_1, r_2) are $\phi_1, \phi_2, \phi_1 + 2\phi_2, \phi_1 - 2\phi_2$. The only angle that is not timelike is $\chi = \phi_1 - 2\phi_2$. Averaging produces the approximating system:

$$\dot{r}_1 = -\frac{\varepsilon}{8}a_2 r_2^2 \cos(\phi_1 - 2\phi_2),\ \dot{r}_2 = \frac{\varepsilon}{2}a_2 r_1 r_2 \cos(\phi_1 - 2\phi_2).$$

If we use the $O(\varepsilon)$ terms of the equations for ϕ_1, ϕ_2 in system (8.10) we obtain after averaging the same results as before: periodic solutions and integrals. The concept of timelike and not timelike angles helps us to obtain a slowly varying equation for χ. Domains in space where $d\chi/dt = 0$ correspond with resonance manifolds; this is where we located periodic solutions.

This calculation would be more complicated if the $O(1)$ variation of the angles ϕ_1, ϕ_2 would depend on r_1, r_2. Near stable equilibrium in phase-space one can often safely assume constant frequencies but not always for larger values of the energy.

We conclude that the advantage of using amplitude-angles variables is that we know that the size of the resonance zone containing the periodic solutions in this case $O(\sqrt{\varepsilon})$. Apart from that, the different variational approaches produce similar results here.

We observe that the Hamiltonian (8.7) is a potential problem with 4 free parameters. If $a_1 = a_2 = 0$ the potential is discrete (mirror) symmetric in variable x. In addition, only one of them, a_2, plays a part in the first order approximation. If $a_2 = 0$ the first order approximation is trivial.

The Case $a_2 = 0$
In this case Hamiltonian (8.7) leads to the equations of motion:

$$\begin{cases} \ddot{x} + 4x & = \varepsilon(a_1 x^2 + 2a_4 xy), \\ \ddot{y} + y & = \varepsilon(a_3 y^2 + a_4 x^2). \end{cases}$$

$$(8.11)$$

As noted above, first order averaging leads to trivial results. Second order averaging leads to:

$$\dot{r}_1 = O(\varepsilon^3),\ \dot{r}_2 = O(\varepsilon^3),$$

$$(8.12)$$

and a nontrivial equation for the combination angle $\chi = 2\phi_1 - 4\phi_2$:

$$\frac{d\chi}{dt} = \varepsilon^2 f(r_1, r_2) + O(\varepsilon^3).$$

In this case we have to consider the potential problem as a 4:2 higher order resonance. The angle χ is timelike except in the resonance manifold determined by $f(r_1, r_2) = 0$.

More explicit details are given in the following physical examples.

8.4.2 A Cubic Potential in Higher Order Resonance

Consider again the cubic potential problem studied in Chap. 2 for the 1:2 resonance and in Chap. 6 for the 1:1 resonance. It represents a strongly simplified galactic potential problem. Consider the Hamiltonian with higher order resonance:

$$H(p, q) = \frac{1}{2}(p_1^2 + k^2 q_1^2) + \frac{1}{2}(p_2^2 + l^2 q_2^2) - \varepsilon q_1 q_2^2. \tag{8.13}$$

The positive frequencies k, l are relative prime and $k + l \geq 5$ The equations of motion are:

$$\begin{cases} \ddot{q}_1 + k^2 q_1 &= \varepsilon q_2^2, \\ \ddot{q}_2 + l^2 q_2 &= 2\varepsilon q_1 q_2. \end{cases} \tag{8.14}$$

Following Remark 1 we define $\chi = l\phi_1 - k\phi_2$. For transformation to slowly varying variables one has several choices. Using the non-canonical but quantitatively equivalent amplitude-phase variables (1.6) we find by averaging to second order:

$$\begin{cases} \dot{r}_1 &= O(\varepsilon^3), \dot{r}_2 = O(\varepsilon^3), \\ \dot{\chi} &= \varepsilon^2 \frac{k^5}{l(k^2 - 4l^2)} \left(\frac{3k^2 - 4l^2}{4k^2} r_2^2 - r_1^2 \right). \end{cases} \tag{8.15}$$

Rescaling $\varepsilon^2 t = \tau$ we recognise the set-up of Theorem 7.1 for averaging over one angle. The resonance manifold, if it exists, is given by the equation:

$$\frac{3k^2 - 4l^2}{4k^2} r_2^2 - r_1^2 = 0. \tag{8.16}$$

As before (Chaps. 2 and 6) we have the approximate energy integral:

$$k^2 r_1^2 + l^2 r_2^2 = 2E_0.$$

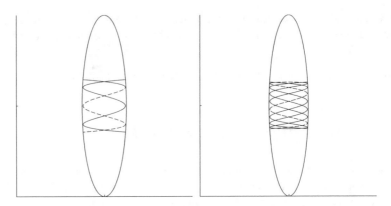

Fig. 8.1 Projections of periodic solutions (stable full line, unstable hashed) on the (q_1, q_2) plane of the $k/l = 4/1$ and $9/2$ resonances described for system (8.14); the figures are based on [57]

Using condition (8.16) and the approximate energy integral we find for the location of the resonance manifold depending on the energy level E_0:

$$r_1^2 = \frac{3k^2 - 4l^2}{3k^4} 2E_0, \quad r_2^2 = \frac{8}{3k^2} E_0. \tag{8.17}$$

The conditions that $k^2 r_1^2$ and $l^2 r_2^2$ have to be smaller than $2E_0$ lead with Eq. (8.17) to the existence condition of this resonance manifold $m/n > 2/\sqrt{3}$. In Fig. 8.1 we show projections of the stable (full line) and unstable (hashed) periodic solutions in 2 cases of m/n.

8.4.3   The Spring-Pendulum in Higher Order Resonance

As a classical example consider again the elastic pendulum, a pendulum where the suspending, inflexible string is replaced by a linear spring. It was introduced in Sect. 2.3.9 for the 1:2 resonance. At rest the spring has length l_0, a mass m is attached to the spring, g is the gravitational constant. The spring has 2 dof as it can oscillate in the vertical direction and swing like a pendulum in one vertical plane. We will look at the *higher order* resonances. As predicted by Theorem 8.1, if higher order resonance is active there will be two domains in phase-space where the dynamics is very different and they are characterised by different timelike variables.

Let $r(t)$ be the length of the spring at time t and ϕ the angular deflection of the spring with respect to the vertical. Neglecting friction the Hamiltonian is:

$$H = \frac{1}{2m} \left(p_r^2 + \frac{p_\phi^2}{r^2} \right) + \frac{s}{2} (r - l_0)^2 - mgr \cos\phi, \tag{8.18}$$

with s a positive constant, $p_r = m\dot{r}$, $p_\phi = mr^2\dot{\phi}$. In [64] a few rescalings are introduced like for the length of the spring $z = (r - l_0)/l_0$, also that the spring-pendulum is at rest if $(p_z, z, p_\phi, \phi) = (0, 0, 0, 0)$. The basic frequencies are $\omega_z = \sqrt{s/m}$, $\omega_\phi = \sqrt{g/l_0}$ producing for the linearised motion in the vertical and horizontal directions respectively:

$$\ddot{z} + \omega_z^2 z = 0, \quad \ddot{\phi} + \omega_\phi^2 \phi = 0.$$

Expanding Hamiltonian (8.18) in the rescaled variables we have:

$$H_2 = \frac{1}{2}\omega_z(p_z^2 + z^2) + \frac{1}{2}\omega_\phi(p_\phi^2 + \phi^2), \tag{8.19}$$

$$H_3 = \frac{\omega_\phi}{\sqrt{ml_0^2\omega_z}}\left(\frac{1}{2}z\phi^2 - zp_\phi^2\right), \tag{8.20}$$

$$H_4 = \frac{1}{ml_0^2}\left(\frac{3}{2}\frac{\omega_\phi}{\omega_z}z^2 p_\phi^2 - \frac{1}{24}\phi^4\right). \tag{8.21}$$

For simplicity we replace ω_z, ω_ϕ by the natural numbers k, l. For averaging we rescale the variables by ε, $z \mapsto \varepsilon\bar{z}$ etc. followed by dividing by ε and omitting the bars.

It was an interesting exercise in Sect. 2.3.9 to put $k = 2, l = 1$ and find that first order averaging using transformation (1.6) produces interesting results arising from the H_3 terms.

Assume now that $k/l \neq 2$ and also not ε-close to this ratio. Applying second order averaging we find

$$\dot{r}_1 = O(\varepsilon^3), \quad \dot{r}_2 = O(\varepsilon^3),$$

and for the combination angle χ, see (8.5), following [64] an expression of the form

$$\dot{\chi} = l\dot{\psi}_1 - k\dot{\psi}_2 = f(r_1, r_2),$$

with f depending on the frequencies k, l. We present the expression in the case of the 6:1 resonance:

$$\dot{\chi} = -\varepsilon^2\left(\frac{4.57}{\sigma}r_1^2 - 0.46r_2^2\right), \tag{8.22}$$

with $\sigma > 0$, so the righthand side has a zero for fixed energy. The 6:1 resonance manifold exists. In Fig. 8.2 we present a Poincaré map for the $6 : 1$ resonance with the resonance domain depicted as a small region $O(\varepsilon^{3/2})$.

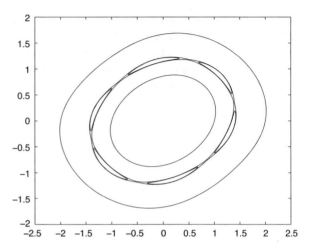

Fig. 8.2 Poincaré map for the 6:1-resonance of the spring-pendulum ($\varepsilon = 0.75$, large for illustration purposes). In the resonance zone, the saddles are connected by heteroclinic cycles and inside the cycles are centres, see [64], courtesy SIAP

Table 8.1 The table from [64] presents the most prominent higher order resonances of the spring-pendulum with lowest order resonant terms $O(\varepsilon^{k+l-2})$. The third column gives the size of the resonance zone in which the resonance manifold M is embedded, while in the fourth column we find the timescale of interaction in the resonance zone

Resonance	$k + l - 2$	d_ε	Interaction timescale
1:4	3	$\varepsilon^{1/2}$	$\varepsilon^{-5/2}$
3:4	5	$\varepsilon^{3/2}$	$\varepsilon^{-7/2}$
1:6	5	$\varepsilon^{3/2}$	$\varepsilon^{-7/2}$
2:6	6	ε^{2}	ε^{-4}
1:8	7	$\varepsilon^{5/2}$	$\varepsilon^{-9/2}$
4:6	8	ε^{3}	ε^{-5}

Following [64] we summarise some results in Table 8.1. It is clear that the larger $k+l$, the smaller the resonance domain indicated by size d_ε. The interaction intervals of time between the 2 dof in the resonance zones increase with $k + l$.

8.4.4 Three dof, the 1:2:1 Resonance

The case of 3 dof Hamiltonian systems involves many new technical problems and qualitative problems. Technical, as the number of terms increases, qualitative as the complexity of periodic solutions and invariant manifolds increases enormously if the number of dof is larger than 2. This is partly illustrated by the integrability of the averaged (or normalised) Hamiltonian systems. A n dof time-independent Hamiltonian is called *integrable* if n independent first integrals exist foliating the bounded $(2n - 1)$-dimensional energy manifold in smooth manifolds on which the solutions are moving. For 2 dof Hamiltonian systems the averaged Hamiltonian system is always integrable near stable equilibrium even if the original system is not integrable. This means that because of the error estimates of the approximate phase-flow possible chaotic behaviour has to be a small-scale affair, $O(\varepsilon)$ or even smaller. As we shall see, the averaged Hamiltonian system with 3 dof is in general not integrable although special symmetries or coefficients in the Hamiltonian may change the picture.

In [68] the 4 first low order resonance cases are presented, in [58] a survey of results is given. As an example we consider the 1:2:1 resonance.

In [68] the general Hamiltonian (8.1) is studied to third degree, $H = H_2 + \varepsilon H_3$. The cubic term contains 56 parameters, after first order averaging 6 parameters. Three combination angles arise in the first order averaged system:

$$\chi_1 = 2\phi_1 - \phi_2 - a_1, \ \chi_2 = \phi_1 - \phi_2 + \phi_3 - a_2, \ \chi_3 = 2\phi_3 - \phi_2 - a_3,$$

with a_1, a_2, a_3 parameters; 3 other parameters are coefficients of the terms in the averaged system. There exist 7 families of periodic solutions parameterised by the energy, only 2 families are stable. The only linear mode that can be continued is the mode associated with frequency 2, it is unstable. Two integrals of motion of the averaged system could be found; later it was shown in [21] that in general a third integral does not exist, the averaged system of the 1:2:1-resonance is in general not integrable. "In general" means here that we have integrability only for specific, isolated parameter values.

It is not easy to visualise the dynamics of systems with many variables, in this case 6, but in some cases we can use global quantities. For Hamiltonian systems we have the Poincaré recurrence theorem 8.2. We will use the Euclidean distance $d(t)$ (8.6) for a three dof system.

To be more specific we consider the slightly simpler general potential problem:

$$H(p, q) = \frac{1}{2}(p_1^2 + q_1^2) + \frac{1}{2}(p_2^2 + 4q_2^2) + \frac{1}{2}(p_3^2 + q_3^2) - \varepsilon H_3, \tag{8.23}$$

with 10 parameters, $H_3 =$

$$\frac{b_1}{3}q_1^3 + b_2q_1^2q_2 + b_3q_1^2q_3 + \frac{b_4}{3}q_2^3 + b_5q_2^2q_1 + b_6q_2^2q_3$$

$$+ \frac{b_7}{3}q_3^3 + b_8q_3^2q_1 + b_9q_3^2q_2 + b_{10}q_1q_2q_3.$$

The equations of motion are:

$$\begin{cases} \ddot{q}_1 + q_1 &= b_1q_1^2 + 2b_2q_1q_2 + 2b_3q_1q_3 + b_5q_2^2 + b_8q_3^2 + b_{10}q_2q_3, \\ \ddot{q}_2 + 4q_2 &= b_2q_1^2 + b_4q_2^2 + 2b_5q_2q_1 + 2b_6q_2q_3 + b_9q_3^2 + b_{10}q_1q_3, \quad (8.24) \\ \ddot{q}_3 + q_3 &= b_3q_1^2 + b_6q_2^2 + b_7q_3^2 + 2b_8q_3q_1 + 2b_9q_3q_2 + b_{10}q_1q_2. \end{cases}$$

Only the parameter b_2, b_9, b_{10} are left. Excluding the normal mode planes we use amplitude-phase variables (1.6) and find with the usual scaling $q_1 \mapsto \varepsilon q_i$ etc. after averaging:

$$
\begin{cases}
\dot{r}_1 = -\varepsilon(\frac{b_2}{2}r_1r_2 \sin(2\phi_1 - \phi_2) + \frac{b_{10}}{4}r_2r_3 \sin(\phi_1 + \phi_3 - \phi_2)), \\
\dot{\phi}_1 = -\varepsilon(\frac{b_2}{2}r_2 \cos(2\phi_1 - \phi_2) + \frac{b_{10}}{4}\frac{r_2r_3}{r_1} \cos(\phi_1 + \phi_3 - \phi_2)), \\
\dot{r}_2 = \frac{\varepsilon}{8}(b_2r_1^2 \sin(2\phi_1 - \phi_2) + b_9r_3^2 \sin(2\phi_3 - \phi_2) + b_{10}r_1r_3 \sin(\phi_1 + \phi_3 - \phi_2)), \\
\dot{\phi}_2 = -\frac{\varepsilon}{8r_2}(b_2r_1^2 \cos(2\phi_1 - \phi_2) + b_9r_3^2 \cos(2\phi_3 - \phi_2) + b_{10}r_1r_3 \cos(\phi_1 + \phi_3 - \phi_2)) \\
\dot{r}_3 = -\varepsilon(\frac{b_9}{2}r_3r_2 \sin(2\phi_3 - \phi_2) + \frac{b_{10}}{4}r_1r_2 \sin(\phi_1 + \phi_3 - \phi_2)), \\
\dot{\phi}_3 = -\varepsilon(\frac{b_9}{2}r_2 \cos(2\phi_1 - \phi_2) + \frac{b_{10}}{4}\frac{r_1r_2}{r_3} \cos(\phi_1 + \phi_3 - \phi_2)).
\end{cases}
$$

$$(8.25)$$

Multiplying the equation for $\dot{r}_1$ by r_1, the equation for $\dot{r}_2$ by $4r_2$, the equation for $\dot{r}_3$ by r_3 and adding the equations, we have after integration that system (8.25) has the integral

$$\frac{1}{2}r_1^2 + 2r_2^2 + \frac{1}{2}r_3^2 = E_0, \tag{8.26}$$

with E_0 a positive constant. We can find periodic solutions in general position by putting the righthand sides of the equations for r_1, r_2, r_3 zero. The analysis is elementary but requires too many details here. In Fig. 8.3 we show the Poincaré recurrence by distance $d(t)$ for 1000 timesteps. The parameters are given in the caption. The instability of the q_2 mode is clear, the recurrence is irregular and weak as on this interval of time the orbit does not approach the initial condition very close.

We will restrict ourselves now to the physically interesting case of symmetries and absence of symmetry with initial values near the q_2 normal mode. Discrete or mirror symmetry is often a typical characteristic in physical models, think of a pendulum that swings in two horizontal directions with respect to a vertical axis. If the horizontal motion corresponds with q_1, q_3 we have the case described as follows:

Mirror Symmetry in q_1 and q_3
Mirror symmetry in q_1, q_3 implies that only b_2, b_4 and b_9 can be unequal to zero. System (8.24) becomes:

$$
\begin{cases}
\ddot{q}_1 + q_1 &= 2b_2q_1q_2, \\
\ddot{q}_2 + 4q_2 &= b_2q_1^2 + b_4q_2^2 + b_9q_3^2, \\
\ddot{q}_3 + q_3 &= 2b_9q_3q_2.
\end{cases}
\tag{8.27}
$$

We have written the system down explicitly to show the difference with the asymmetric system (8.24): the q_2 normal mode exists in system (8.27), if q_1 vanishes we have a q_2, q_3, invariant manifold, if q_3 vanishes we have a q_1, q_2 invariant

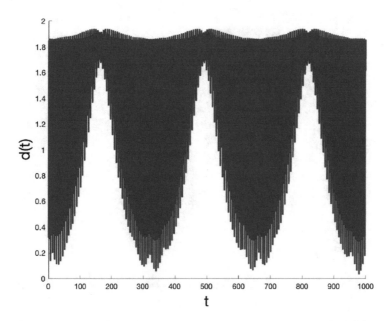

Fig. 8.3 Poincaré recurrence distance $d(t)$ of system (8.24) with initial values $q_1(0) = 0.1, q_2(0) = 0.8, q_3(0) = -0.1$ and initial velocities zero, so near the q_2 normal mode. Parameters: $\varepsilon = 0.05$;, for $b_1, \ldots, b_{10}$ we have $0.2, 0.5, 0.2, 0.5, -0.03, 0.2, -0.4, 0.2, 0.6, -1, 0 \le t \le 1000$

manifold. The averaged system (8.25) reduces to:

$$
\begin{cases}
\dot{r}_1 = -\varepsilon \frac{b_2}{2} r_1 r_2 \sin(2\phi_1 - \phi_2), \dot{\phi}_1 = -\varepsilon \frac{b_2}{2} r_2 \cos(2\phi_1 - \phi_2), \\
\dot{r}_2 = \frac{\varepsilon}{8}(b_2 r_1^2 \sin(2\phi_1 - \phi_2) + b_9 r_3^2 \sin(2\phi_3 - \phi_2)), \\
\dot{\phi}_2 = -\frac{\varepsilon}{8r_2}(b_2 r_1^2 \cos(2\phi_1 - \phi_2) + b_9 r_3^2 \cos(2\phi_3 - \phi_2)), \\
\dot{r}_3 = -\varepsilon \frac{b_9}{2} r_3 r_2 \sin(2\phi_3 - \phi_2), \dot{\phi}_3 = -\varepsilon \frac{b_9}{2} r_2 \cos(2\phi_1 - \phi_2).
\end{cases}
\tag{8.28}
$$

System (8.28) has the integral (8.26). For the 2 combination angles $\chi_1 = 2\phi_1 - \phi_2$, $\chi_3 = 2\phi_3 - \phi_2$ we have:

$$
\frac{d\chi_1}{dt} = -\frac{\varepsilon}{8r_2}\left(b_2(8r_2^2 - r_1^2)\cos\chi_1 - b_9 r_3^2 \cos\chi_3\right),
$$

$$
\frac{d\chi_3}{dt} = -\frac{\varepsilon}{8r_2}\left(b_9(8r_2^2 - r_3^2)\cos\chi_3 - b_2 r_1^2 \cos\chi_1\right).
\tag{8.29}
$$

The amplitudes are constant if $\chi_1 = 0, \pi$ and $\chi_3 = 0, \pi$ producing $\cos\chi_1 = \pm 1, \cos\chi_3 = \pm 1$. This produces 4 possibilities of families of periodic solutions.

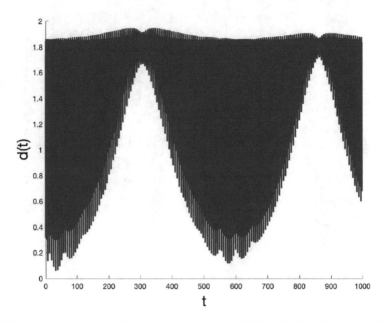

Fig. 8.4 Poincaré recurrence distance $d(t)$ of system (8.27) with initial values $q_1(0) = 0.1, q_2(0) = 0.8, q_3(0) = -0.1$ and initial velocities zero; $\varepsilon = 0.05, b_2 = 0.5, b_4 = 0.5, b_9 = 0.6$ and other b-parameters zero (mirror symmetry in the q_1, q_3 modes). Instability of the q_2 normal mode remains, the recurrence has not improved

For instance if we have $\cos \chi_1 = \cos \chi_3 = 1$ we have from the equations for χ_1, χ_3 the conditions for positive amplitudes:

$$b_2(8r_2^2 - r_1^2) - b_9 r_3^2 = 0, \ b_9(8r_2^2 - r_3^2) - b_2 r_3^2 = 0 \qquad (8.30)$$

Together with integral (8.26) we will look for positive solutions of the amplitudes. With this procedure we find at most 4 families of periodic solutions. We will leave the details to the reader but we illustrate the complexity in Fig. 8.4 by using the recurrence theorem. We choose $b_2 = 0.5, b_4 = 0.5, b_9 = 0.6$; the recurrence is not very good on this interval of time, see Fig. 8.4.

Mirror Symmetry in q_2
In this case we have $b_2 = b_4 = b_9 = b_{10} = 0$ in system (8.24). The system becomes:

$$\begin{cases} \ddot{q}_1 + q_1 &= b_1 q_1^2 + 2b_3 q_1 q_3 + b_5 q_2^2 + b_8 q_3^2, \\ \ddot{q}_2 + 4q_2 &= 2b_5 q_2 q_1 + 2b_6 q_2 q_3, \\ \ddot{q}_3 + q_3 &= b_3 q_1^2 + b_6 q_2^2 + b_7 q_3^2 + 2b_8 q_3 q_1. \end{cases} \qquad (8.31)$$

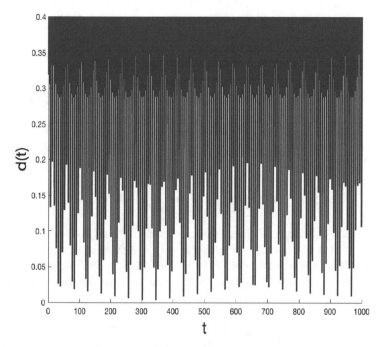

Fig. 8.5 Poincaré recurrence distance $d(t)$ of system (8.27) with initial values $q_1(0) = 0.1, q_2(0) = 0.8, q_3(0) = -0.1$ and initial velocities zero; $\varepsilon = 0.05, b_1 = 0.2, b_3 = 0.2, b_5 = -0.03, b_6 = 0.2, b_7 = -0.4$ and other b-parameters zero (mirror symmetry in the q_2 mode). The recurrence is much faster

The first order averaging produces zero righthand sides, we have to average to second order in this case. This is left to the reader. The phase-flow is to a larger extent determined by the linearised system, the recurrence has strongly improved, see Fig. 8.5.

8.4.5 The Fermi-Pasta-Ulam Chain

In the original Fermi-Pasta-Ulam (FPU) problem one considered a chain of N so-called mono-atomic elements, i.e. all masses equal. We call this the classical FPU-chain and put for the masses $m_1 = m_2 = \ldots = m_N = 1$. The idea of the authors was to demonstrate numerically the statistical mechanics phenomenon of equipartition of energy. Starting with initial values close to one mode of the system, one expected the energy to spread permanently in the system. This contradicts the recurrence theorem at low energy states but it might be possible that at chaotic, larger energy states the recurrence would take extremely long times. However, the authors found

Fig. 8.6 A Fermi-Pasta-Ulam chain of oscillators with fixed endpoints

recurrence numerically for 16 and 32 particles. For the historical background see [27].

The formulation of the FPU-chain runs as follows. We consider a N dof Hamiltonian near stable equilibrium $p = q = 0$, and use a potential V of the ε-rescaled form:

$$V(z) = \frac{1}{2}z^2 + \varepsilon \frac{1}{3}\alpha z^3 + \varepsilon^2 \frac{1}{4}\beta z^4.$$

So, as usual, ε indicates an estimate of the distance to the origin. We speak of an α-chain if $\alpha \neq 0$, $\beta = 0$ and of a β-chain if $\alpha = 0$, $\beta \neq 0$. Apart from α- or β-chains we have two possible models: the FPU-chain with fixed endpoints, see Fig. 8.6 and the periodic chain where the first mass and the last mass are identified. We have for the Hamiltonian:

$$H = \sum_{i=1}^{N} \left(\frac{1}{2}p_i^2 + V(q_{i+1} - q_i) \right).$$
(8.32)

The Fixed Endpoints FPU-Chain with 3 Masses

It is instructive to consider FPU chains with few masses. In [27] the case with 3 masses is discussed for fixed endpoints and α-chains. The Hamiltonian is (notation [27]):

$$H = \sum_{i=0}^{2} \left(\frac{1}{2}(p_i^2 + (q_{i+1} - q_i)^2) + \varepsilon \frac{1}{3}\alpha(q_{i+1} - q_i)^3 \right).$$
(8.33)

The linearised system is not in quasi-harmonic form like for instance system (8.8) (coupled harmonic equations). The eigenvalues produce the frequencies $(3, 3, 0)$. By a linear transformation we can obtain quasi-harmonic equations with from [27] the equivalent Hamiltonian:

$$H = \frac{1}{2}(P_1^2 + P_2^2 + P_3^2 + 3Q_2^2 + 3Q_3^2) + \varepsilon \frac{3\alpha}{\sqrt{2}}(Q_2 Q_3^2 - \frac{1}{3}Q_3^3).$$
(8.34)

The P, Q coordinates were obtained by the diagonalising transformation. As mentioned in [27] Hamiltonian (8.34) is equivalent by rescaling coordinates to the Hénon-Heiles Hamiltonian presented in [35], also discussed in [74]. The equations of motion are then of the form:

$$\ddot{x} + x = -\varepsilon 2xy, \quad \ddot{y} + y = \varepsilon(-x^2 + y^2).$$
(8.35)

The Hénon-Heiles Hamiltonian was chosen in [35] to illustrate the motion of stars in an axi-symmetric galaxy near the plane of symmetry. This brings to mind what Henri Poincaré wrote in an essay: "Mathematics is the art of giving the same name to different things".

The Hénon-Heiles Hamiltonian became famous as it was the first Hamiltonian system where the transition from regular to chaotic motion was demonstrated numerically. If ε is small, say 0.05, we have dominating regular dynamics but increasing the energy, say to 1/8 or 1/6, chaos appears. It is remarkable that we have already such complex behaviour in the case of 3 masses.

In Fig. 8.7 we show at low energy various recurrence behaviour for initially close orbits. We have in the 3 cases $\varepsilon = 0.05$ and initially $x(0) = 0.05$, $\dot{x}(0) = 0$, $\dot{y}(0) = 0$. For the 4th coordinate we have $y(0) = 0.97, 0.1, 0.101$. The complexity is caused by the presence of very small resonance zones between the large scale tori predicted by KAM-theory, see [18] for the general theory. There will be many small resonance zones like this with qualitative different behaviour of orbits.

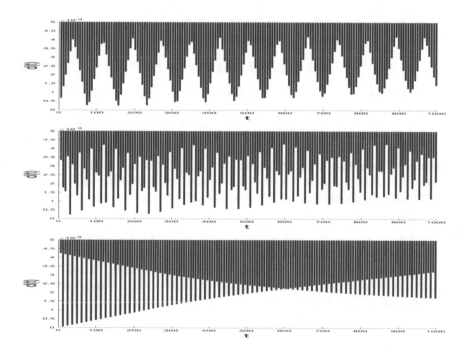

Fig. 8.7 Recurrence of 3 initially vary close orbits of the Fermi-Pasta-Ulam chain of FPU-α coupled oscillators with fixed endpoints where system (8.35) was used on 1000 timesteps; relative and absolute precision e^{-15}. We have $\varepsilon = 0.05$ and plotted $0 \leq d(t) \leq 0.005$ (so 1 on the vertical axis means 0.001). We have initially for the 3 orbits $x(0) = 0.05$, $\dot{x}(0) = 0$, $\dot{y}(0) = 0$. Top figure $y(0) = 0.097$, middle 0.1 and below 0.101. Small differences in $y(0)$ produce large recurrence difference

The Periodic FPU-Chain with 6 Masses

We will consider as a second example the periodic β-chain with 6 particles but as the analysis is rather unwieldy, although mathematically not difficult, we will summarise the results.

We choose $N = 6$. The basic frequencies follow from the linearised part of the equations of motion (leaving out the cubic terms):

$$\begin{cases} \ddot{q}_1 + 2q_1 - q_2 - q_6 = 0, \\ \ddot{q}_2 + 2q_2 - q_3 - q_1 = 0, \\ \ddot{q}_3 + 2q_3 - q_4 - q_2 = 0, \\ \ddot{q}_4 + 2q_4 - q_5 - q_3 = 0, \\ \ddot{q}_5 + 2q_5 - q_6 - q_4 = 0, \\ \ddot{q}_6 + 2q_6 - q_1 - q_5 = 0. \end{cases} \tag{8.36}$$

The linear part of the equations is again not in quasi-harmonic form and we use a diagonalising transformation by MATHEMATICA. Indicate the transformation by $(p, q) \mapsto \bar{p}, \bar{q}$. One can check that the complete nonlinear equations of motion of the periodic chains have the first (momentum) integral:

$$\dot{q}_1 + \dot{q}_2 + \dot{q}_3 + \dot{q}_4 + \dot{q}_5 + \dot{q}_6 = \text{constant}. \tag{8.37}$$

For the transformation we need the eigenvalues and the frequencies $\omega_1, \omega_2, \ldots, \omega_6$ produced by the linear part:

$$1, \sqrt{3}, 2, \sqrt{3}, 1, 0. \tag{8.38}$$

Frequency 0 corresponds with the momentum integral that reduces by transformation the equations of motion to 5. The transformation mixes also the nonlinear terms (this is the unwieldy part that we leave out, see [52]). We introduce the actions:

$$I_i = \frac{1}{2\omega_i}(\bar{p}^2 + \omega_i^2 \bar{q}_i^2), \ i = 1, \ldots, 5. \tag{8.39}$$

It is remarkable that the frequencies contain two 1:1 resonances (1, 1 and $\sqrt{3}, \sqrt{3}$) and one 3 dof resonance (1, 2, 1). Applying averaging-normalisation we find the normalised Hamiltonian $\tilde{H}$ in $I_i, \phi_i, i = 1, \ldots, 5$ variables with:

$$\begin{cases} \tilde{H} = I_1 + \sqrt{3}I_2 + 2I_3 + \sqrt{3}I_4 + I_5 + \\ \quad \varepsilon^2 \beta[\frac{1}{4}\sqrt{3}\sqrt{I_1 I_2 I_4 I_5}(\cos(\phi_1 + \phi_2 - \phi_4 - \phi_5) + \cos(\phi_1 - \phi_2 + \phi_4 - \phi_5)) \\ \quad +\frac{1}{16}I_1 I_5(\cos(2\phi_1 - 2\phi_5) - 1) + \frac{3}{16}I_2 I_4(\cos(2\phi_2 - 2\phi_4) - 1) \\ \quad +\frac{1}{8}\sqrt{3}(I_1 I_2 + I_4 I_5 + 3I_1 I_4 + 3I_2 I_5) + \frac{1}{2}I_3(I_1 + \sqrt{3}I_2 + \sqrt{3}I_4 + I_5) \\ \quad +\frac{1}{4}I_3^2 + \frac{3}{32}(I_1 + I_5)^2 + \frac{9}{32}(I_2 + I_4)^2]. \end{cases} \tag{8.40}$$

From the Hamiltonian $\tilde{H}$ we can write down the equations of motion in actions and angles I, ϕ. As we can see from the Hamiltonian the two 1:1 resonances play a part but the 1:2:1 resonance not. This shows there is a hidden symmetry in the periodic FPU-chain.

In [52] many other aspects are discussed. We mention that there exist 5 independent first integrals of the equations of motion derived from Hamiltonian (8.40). Also, a number of invariant manifolds and periodic solutions for fixed values of the energy can be found.

The recurrence of the periodic FPU β-chain with 6 masses leading to integrable averaged equations of motion is obvious. A remarkable paper [53] shows that because of symmetries and resonances the periodic FPU-chain with arbitrary number of N masses shows the same behaviour. For FPU systems near stable equilibrium the averaged-normalised Hamiltonian is integrable with consequence that recurrence in such systems is dominant. For periodic β-chains there will be no permanent equipartition of energy.

8.4.6 Interaction of Low and Higher Order, the 2:2:3 Resonance

Considering 3 or more dof may present a mix of dominant (low order) resonance and higher order resonance. The Hamiltonian 2:2:3 resonance contains the primary resonance 1:1 and twice the higher order resonance 2:3. We present some aspects of this case based on [81]. Consider for the 2:2:3 resonance the Hamiltonian $H_2 + \varepsilon^2 H_4 + \varepsilon^3 H_5$:

$$
\begin{cases}
H = \frac{1}{2}(\dot{q}_1^2 + 4q_1^2) + \frac{1}{2}(\dot{q}_2^2 + 4q_2^2) + \frac{1}{2}(\dot{q}_3^2 + 9q_3^2) \\
\quad - \frac{1}{4}\varepsilon^2(\alpha_1 q_1^4 + 2\alpha_2 q_1^2 q_2^2 + \alpha_3 q_2^4 + \alpha_4 q_3^4) \\
\quad - \varepsilon^3(b_1 q_1^3 q_3^2 + b_3 q_2^3 q_3^2 + b_3 q_1^2 q_2 q_3^2 + b_4 q_1 q_2^2 q_3^2).
\end{cases} \tag{8.41}
$$

H_4 in the Hamiltonian was chosen to be discrete symmetric in the positions and to contain one combination angle at H_4 level: ($\chi_1 = \phi_1 - \phi_2$) and 4 combination angles at H_5 level. There are many problems in physics with such a slight breaking of symmetry. The equations of motion induced by (8.41) are:

$$
\begin{cases}
\ddot{q}_1 + 4q_1 = \varepsilon^2(\alpha_1 q_1^3 + \alpha_2 q_1 q_2^2) + \varepsilon^3(3b_1 q_1^2 q_3^2 + 2b_3 q_1 q_2 q_3^2 + b_4 q_2^2 q_3^2), \\
\ddot{q}_2 + 4q_2 = \varepsilon^2(\alpha_2 q_1^2 q_2 + \alpha_3 q_2^3) + \varepsilon^3(3b_2 q_2^2 q_3^2 + b_3 q_1^2 q_3^2 + 2b_4 q_1 q_2 q_3^2), \\
\ddot{q}_3 + 9q_3 = \varepsilon^2 \alpha_4 q_3^3 + \varepsilon^3(2b_1 q_1^3 q_3 + 2b_2 q_2^3 q_3 + 2b_3 q_1^2 q_2 q_3 + 2b_4 q_1 q_2^2 q_3).
\end{cases}
$$
$$
\tag{8.42}
$$

The figures we show in this section have been obtained using system (8.42). To make some explicit numerical calculations we have in most cases chosen:

$$\alpha_1 = 0.4, \alpha_2 = 1, \alpha_3 = 0.6, \alpha_4 = 4, b_1 = 1, b_2 = -1.5, b_3 = 1, b_4 = -1. \tag{8.43}$$

The choice of $\alpha_1, \ldots, \alpha_4$ excludes non-generic behaviour. Because of the resonances in H_2 and the resulting frequency ratio's we expect resonant interaction between the first two modes q_1, q_2 and a small interaction with the third mode.

The Primary Resonance Zones
At H_5 four combination angles are possibly active:

$$\chi_2 = 3\phi_1 - 2\phi_3, \ \chi_3 = 3\phi_2 - 2\phi_3, \ \chi_4 = 2\phi_1 + \phi_2 - 2\phi_3, \ \chi_5 = \phi_1 + 2\phi_2 - 2\phi_3.$$

We will use amplitude-phase coordinates r, ϕ from (1.6). The transformation results in:

$$H_2 = 2r_1^2 + 2r_2^2 + \frac{9}{2}r_3^2.$$

We find after averaging to $O(\varepsilon^2)$:

$$\begin{cases} \dot{r}_1 = -\varepsilon^2 \frac{1}{8} r_1 r_2^2 \sin 2\chi_1, \ \dot{\phi}_1 = -\varepsilon^2 \frac{1}{8} (\frac{6}{5} r_1^2 + 2r_2^2 + r_2^2 \cos 2\chi_1), \\ \dot{r}_2 = +\varepsilon^2 \frac{1}{8} r_1^2 r_2 \sin 2\chi_1, \ \dot{\phi}_2 = -\varepsilon^2 \frac{1}{8} (2r_1^2 + r_1^2 \cos 2\chi_1 + \frac{9}{5} r_2^2), \quad (8.44) \\ \dot{r}_3 = 0, \ \dot{\phi}_3 = -\varepsilon^2 \frac{3}{8} r_3^2. \end{cases}$$

The combination angles $\chi_2, \ldots, \chi_5$ will arise at $O(\varepsilon^3)$. For the averaging-normal form (8.44) we have $2r_1^2 + 2r_2^2 = E_1 + O(\varepsilon)$ with constant $E_1 \geq 0$ and $r_3 - r_3(0) = O(\varepsilon)$ on the timescale $1/\varepsilon^2$. The normal form to $O(\varepsilon^2)$ is integrable with integrals $r_3 = r_3(0), 2r_1^2 + 2r_2^2 = E_1$ and the normalised Hamiltonian to quartic terms.

The equation for χ_1 depends on r_1, r_2 and χ_1. If the averaging-normal form (8.44) of the equations has a few solutions with $\sin 2\chi_1 = 0, \dot{\chi}_1 = 0$, they give approximate short-periodic solutions in general position (separate from the normal modes).

A *primary resonance zone* M is defined as a neighbourhood of the solutions with $\sin 2\chi_1 = 0, \dot{\chi}_1 = 0$. In the 2:2:3 problem such a resonance zone has for a chosen value E_0 with $H_2 = E_0, E_0 > 0$ the free parameter $r_3(0)$. We find with system (8.44):

$$\dot{\chi}_1 = \frac{1}{8}\varepsilon^2 \left(\frac{4}{5}r_1^2 - \frac{1}{5}r_2^2 + (r_1^2 - r_2^2)\cos 2\chi_1 \right). \tag{8.45}$$

Putting the righthand side of Eq. (8.45) zero we find two primary resonance zones M_1, M_2:

$$\begin{cases} M_1 : \ 2\chi_1 = \ \ 0, 2\pi, r_1^2 = \frac{2}{3}r_2^2, \\ M_2 : \ 2\chi_1 = \ \ \pi, 3\pi, r_1^2 = 4r_2^2. \end{cases} \tag{8.46}$$

The periodic solutions of the two primary resonance zones are located on the energy manifold which is topologically the 5-dimensional sphere S^5 and its intersections with the elliptical cylinder $\frac{1}{2}(\dot{q}_1^2 + 4q_1^2) + \frac{1}{2}(\dot{q}_2^2 + 4q_2^2) = E_1$ and the hyperbolic cylinder $\dot{q}_1^2 + 4q_1^2 = 4r_1^2 = \mu(\dot{q}_2^2 + 4q_2^2)$ with respectively $\mu = \frac{2}{3}$ or 4; this defines two three-dimensional manifolds.

To study the complexity of the primary resonance zones we use the recurrence theorem 8.2. In Fig. 8.8 we start outside the resonance zones to observe the resonant interaction of the first two modes; the variations of the third mode are small in accordance with the normal form (8.44). We will use the Euclidean distance $d(t)$ for three dof of Eq. (8.6). Choosing $q_3(0) = \dot{q}_3(0) = 0$ the third mode remains zero, the recurrence is fairly strong and regular; see Fig. 8.8 right. Starting with nonzero initial values of the third mode in Fig. 8.8 left we find behaviour that suggests quasi-trapping in one or more resonance zones (see [84]). This suggests that we have to study the resonance zones in more detail (Fig. 8.9).

We will extend the theory of higher order resonance, see [58], to three dof. Higher order (or secondary) resonance may take place in a primary resonance zone.

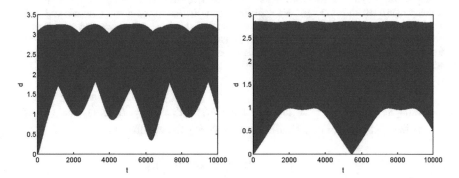

Fig. 8.8 The 2:2:3 resonance. Left the Euclidean distance d using Eq. (8.6) in 10,000 timesteps of the orbits to their initial conditions in phase-space outside the primary resonance zones and parameter values Eq. (8.43); recurrence takes more than 10,000 timesteps. Right the Euclidean distance d for the same Hamiltonian (8.41) but starting at $q_1(0) = 0.3, q_2(0) = 1.2, q_3(0) = 0$ and velocities zero; we have now $q_3(t) = \dot{q}_3(t) = 0, t \geq 0$. The recurrence is different and faster in this case of pure q_1, q_2 interaction. Courtesy IJBC, [81]

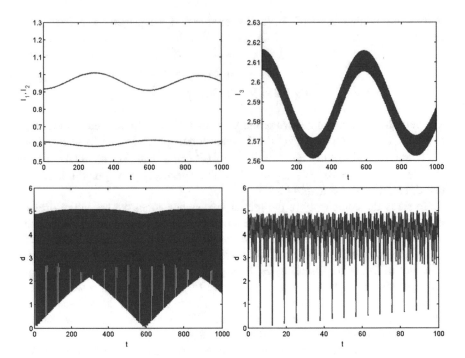

Fig. 8.9 Recurrence when starting in the primary resonance zone M_1 in 1000 timesteps. $E_1 = 3.06$ as in Fig. 8.8 with now $q_1(0) = 0.7823, q_2(0) = 0.9581; q_3(0) = 1.32068$, the value given by Eq. (8.48) which puts the orbits in the secondary resonance zone. Initial velocities are zero. The top figure left shows the variations of the actions I_1, I_2, the top figure right shows I_3 with small variations as predicted. The recurrence distance d (below left) is strong and regular as the orbits are caught in the resonance zone M_1 with near fairly stable dynamics; the picture for 1000 timesteps shows some white segments but hides the fine-structure of recurrence transitions shown when enlarged for 100 timesteps (below right)

Consider as an example the primary zone M_1 with $2\chi_1 = 0, 2\pi, r_1^2 = \frac{2}{3}r_2^2$. To $O(\varepsilon^2)$ the higher order combination angles satisfy the equations:

$$\begin{cases} \dot{\chi}_2 = & -\frac{3}{4}\varepsilon^2(\frac{3}{5}r_1^2 + \frac{3}{2}r_2^2 - r_3^2), \\ \dot{\chi}_3 = & -\frac{3}{4}\varepsilon^2(\frac{3}{2}r_1^2 + \frac{9}{10}r_2^2 - r_3^2), \\ \dot{\chi}_4 = & -\frac{3}{4}\varepsilon^2(\frac{9}{10}r_1^2 + \frac{13}{10}r_2^2 - r_3^2), \\ \dot{\chi}_5 = & -\frac{3}{4}\varepsilon^2(\frac{6}{5}r_1^2 + \frac{11}{10}r_2^2 - r_3^2). \end{cases} \qquad (8.47)$$

Outside the primary resonance zone we have terms in system (8.44) from the first two modes that dominate at $O(\varepsilon^2)$. In the zone M_1 this is not the case and other combination angles may play a part. If the derivative of a combination angle is sign-definite, the angle is timelike and we can average over the angle. A combination angle is not timelike if the righthand side of the equation contains zeros. In such a case a secondary resonance will be found in the primary resonance zone. In M_1 we find for the 1:1 periodic solutions:

$$2r_1^2 + 2r_2^2 = E_1, \ r_1^2 = \frac{2}{3}r_2^2 \ \rightarrow \ r_1^2 = \frac{1}{5}E_1, r_2^2 = \frac{3}{10}E_1.$$

The combination angles in (8.47) admit a zero (critical value) if in M_1:

$$r_3^2 = \frac{57}{100}E_1. \tag{8.48}$$

We repeat the calculation for primary resonance zone M_2. We find for the 1:1 periodic solutions in M_2:

$$2r_1^2 + 2r_2^2 = E_1, \ r_1^2 = 4r_2^2 \ \rightarrow \ r_1^2 = \frac{2}{5}E_1, r_2^2 = \frac{1}{10}E_1.$$

The combination angles in (8.47) (we omit the details) admit a zero (critical value) if in M_2:

$$r_3^2 = \frac{29}{100}E_1. \tag{8.49}$$

To improve our insight in the dynamics we produce the actions corresponding with the two cases of Fig. 8.10. In Fig. 8.11 we present left I_1 and I_2 starting outside the primary resonance zones exchanging energy according to the 1:1 resonance; I_3 (right) varies within the error estimates. Next we give the actions in the unstable

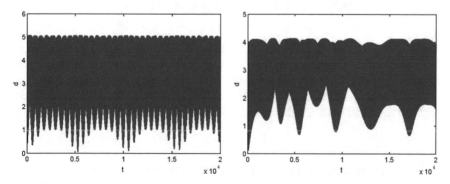

Fig. 8.10 The 2:2:3 resonance. Recurrence when starting outside primary resonance zone M_1 and M_2 in 20,000 timesteps. $E_1 = 3.06$ as in Fig. 8.8 with now left $q_1(0) = 0.3, q_2(0) = 1.2$; $q_3(0) = 1.32068$, the critical value in M_1 given by Eq. (8.48). The orbits passing through M_1 encounter the critical case of $q_3(0)$, recurrence is slow and different from the case of Fig. 8.8. Right the case $q_1(0) = 0.3, q_2(0) = 1.2$; $q_3(0) = 0.9420$, the critical value in M_2 given by Eq. (8.49). The recurrence takes much longer which suggests that this location corresponds with unstable periodic solutions

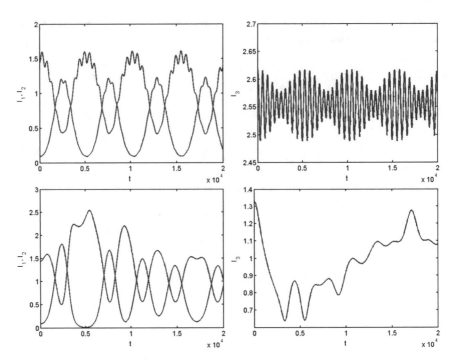

Fig. 8.11 The 2:2:3 resonance. The actions when starting outside primary resonance zone M_1 and M_2 in 20,000 timesteps. Each orbit starts outside the primary resonance zones with $q_1(0) = 0.3$, $q_2(0) = 1.2$ and all velocities zero. In the first two figures (top) we took $q_3(0) = 1.32068$, the critical value in M_1 given by Eq. (8.48). $I_3(t)$ varies with magnitude 0.06 in according with the error estimate. The instability of the normal modes forces considerable exchange of energy of the first two modes. The next two figures show the instability of M_2. When passing this primary resonance zone the critical value of $q_3(0) = 0.9420$ taken from Eq. (8.49) plays a part

case of Fig. 8.10 (right). It is interesting that $I_3(t)$ changes strongly (between 0.65 and 1.33) when outside the time-interval of validity, $1/\varepsilon^2$, of the normal form. Conservation of energy influences the exchanges of I_1 and I_2 at the same time.

To obtain a better understanding of the dynamics in the primary resonance zones we have to extend the averaging-normal forms to $O(\varepsilon^3)$ (Fig. 8.12).

8.4.7 Interaction of Low and Higher Order, the 1:1:4 Resonance

Embedded double resonances can be discussed in the framework of interaction between lower and higher order Hamiltonian resonances of type 1:1:ω. After the

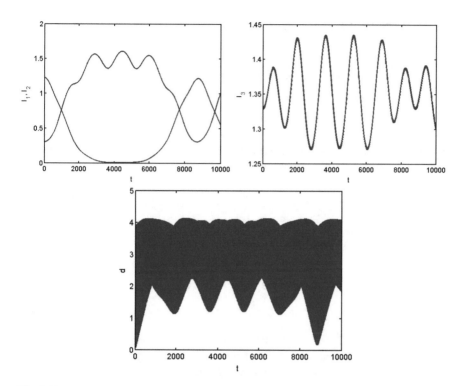

Fig. 8.12 The 2:2:3 resonance. Dynamics when starting in the primary resonance zone M_2 in (because of the instability) 10,000 timesteps. $E_1 = 3.06$ as in Fig. 8.8 with now $q_1(0) = 1.1063, q_2(0) = 0.5532; q_3(0) = 0.94202$, values putting the orbits in M_2. Initial velocities are zero. The top figure left shows strong variations of the actions I_1, I_2, the top figure right shows I_3; the variations are strong as we start near an unstable secondary resonance. The orbits are leaving the unstable resonance zone M_2. Below the corresponding recurrence d from Eq. (8.6)

resonance 2:2:3 we present another example here showing interesting stability aspects. We illustrate the dynamics of the 1:1:4 resonance for the Hamiltonian:

$$
\begin{cases}
H &= \frac{1}{2}(\dot{q}_1^2 + q_1^2) + \frac{1}{2}(\dot{q}_2^2 + q_2^2) + \frac{1}{2}(\dot{q}_3^2 + 16q_3^2) \\
&\quad -\frac{1}{4}\varepsilon^2(a_1 q_1^4 + 2a_2 q_1^2 q_2^2 + a_3 q_2^4 + a_4 q_3^4) \\
&\quad -\varepsilon^3(b_1 q_1^4 q_3 + b_2 q_2^4 q_3).
\end{cases}
\tag{8.50}
$$

Note that H_3 is discrete (mirror) symmetric in the 3 positions, H_4 is discrete symmetric in q_1, q_2. Such symmetries are often found in physics applications. We define the actions $I_i = \frac{1}{2}(\dot{q}_i^2 + q_i^2), i = 1, 2$ and $I_3 = \frac{1}{2}(\dot{q}_3^2 + 16q_3^2)$. The Hamiltonian was chosen with the cubic terms already transformed to higher order by near-identity transformation, see [58] ch. 10. The equations of motion induced by (8.50) are:

$$\begin{cases} \ddot{q}_1 + q_1 & = \varepsilon^2 (a_1 q_1^3 + a_2 q_1 q_2^2) + \varepsilon^3 4 b_1 q_1^3 q_3, \\ \ddot{q}_2 + q_2 & = \varepsilon^2 (a_2 q_1^2 q_2 + a_3 q_2^3) + \varepsilon^3 4 b_2 q_2^3 q_3, \\ \ddot{q}_3 + 16 q_3 & = \varepsilon^2 a_4 q_3^3 + \varepsilon^3 (b_1 q_1^4 + b_2 q_2^4). \end{cases} \tag{8.51}$$

We will use again polar (amplitude-phase) coordinates r, ψ with transformation (1.6); we find after averaging to $O(\varepsilon^2)$:

$$\begin{cases} \dot{r}_1 = -\varepsilon^2 \frac{a_2}{8} r_1 r_2^2 \sin 2\chi_1 + \varepsilon^3 \dots, \\ \dot{\psi}_1 = -\varepsilon^2 \frac{1}{8} (3 a_1 r_1^2 + 2 a_2 r_2^2 + a_2 r_2^2 \cos 2\chi_1) + \varepsilon^3 \dots, \\ \dot{r}_2 = \varepsilon^2 \frac{a_2}{8} r_1^2 r_2 \sin 2\chi_1 + \varepsilon^3 \dots, \\ \dot{\psi}_2 = -\varepsilon^2 \frac{1}{8} (2 a_2 r_1^2 + a_2 r_1^2 \cos 2\chi_1 + 3 a_3 r_2^2) + \varepsilon^3 \dots, \\ \dot{r}_3 = \varepsilon^3 \dots, \quad \dot{\psi}_3 = -\varepsilon^2 \frac{3 a_4}{32} r_3^2 + \varepsilon^3 \dots. \end{cases}$$

$$(8.52)$$

with $\chi_1 = \psi_1 - \psi_2$. The dots represent higher order terms containing the amplitudes and, as we shall see, the combination angles $\chi_2 = 4\phi_1 - \phi_3$, $\chi_3 = 4\phi_2 - \phi_3$. For the averaging-normal form (8.52) we have the integrals:

$$r_1^2 + r_2^2 = 2E_1, \quad r_3 = r(0)$$

(E_1 constant) with both integrals $O(\varepsilon)$ valid on the timescale $1/\varepsilon^2$.

As before, a primary resonance zone M is defined as a neighbourhood of the solutions of $\sin 2\chi_1 = 0$, $\dot{\chi}_1 = 0$. We find with system (8.52):

$$\dot{\chi}_1 = -\frac{1}{8} \varepsilon^2 \left((3 a_1 - 2 a_2) r_1^2 + (2 a_2 - 3 a_3) r_2^2 + a_2 (r_2^2 - r_1^2) \cos 2\chi_1 \right). \tag{8.53}$$

Putting the righthand side of Eq. (8.53) zero we find the primary resonance zones M_1, M_2:

$$\begin{cases} M_1 : r_1^2 = \frac{a_2 - a_3}{a_2 - a_1} r_2^2, \\ M_2 : r_1^2 = \frac{a_2 - 3 a_3}{a_2 - 3 a_1} r_2^2. \end{cases} \tag{8.54}$$

We require the righthand sides of r_1^2 in (8.54) to be finite and positive. It is easy to check analytically and numerically that outside the primary resonance zones defined by (8.54) we find 1:1 resonant interaction of the first two modes and small modulation of the third mode.

Secondary Resonance in the Primary Resonance Zones

For the combination angles χ_2, χ_3, we find with $\sin 2\chi_1 = 0$ in the primary resonance zones:

$$\begin{cases} \dot{\chi}_2 = -\frac{1}{8}\varepsilon^2(12a_1r_1^2 + 8a_2r_2^2 + 4a_2r_2^2\cos 2\chi_1 - \frac{3}{4}a_4r_3^2), \\ \dot{\chi}_3 = -\frac{1}{8}\varepsilon^2(8a_2r_1^2 + 12a_3r_2^2 + 4a_2r_1^2\cos 2\chi_1 - \frac{3}{4}a_4r_3^2). \end{cases} \tag{8.55}$$

Secondary resonances involving the combination angles in M_1 and M_2 may arise where the righthand side of system (8.55) vanishes; in such a case the combination angles χ_2, χ_3 are not timelike. To study the stability within the resonance zones we compute $\dot{\phi}_3$ to $O(\varepsilon^3)$. We find:

$$\dot{\phi}_3 = -\varepsilon^2\frac{3\alpha_4}{32}r_3^2 - \varepsilon^3\frac{69}{128^2}\alpha_4^2r_3^4 - \frac{\varepsilon^3}{64r_3}A(r_1, r_2, \chi_2, \chi_3)$$

with

$$A(r_1, r_2, \chi_2, \ldots, \chi_6) = b_1r_1^4\cos \chi_2 + b_2r_2^4\cos \chi_3.$$

We conclude that unless the coefficients b_1, b_2, α_4 vanish, $\dot{\phi}_3$ may move out of resonance in the primary resonance zones on a timescale longer than $1/\varepsilon^2$. The higher order resonances included in $M_1, \ldots, M_4$ may generate instabilities, see Fig. 8.13. If secondary resonances arise from zeros of $\dot{\chi}_2, \dot{\chi}_3$, we find stable and unstable periodic solutions from the normal forms in the resonance zones.

For illustrations we postponed the choice of the constants $\alpha_1, \ldots, \alpha_4$ to be able to exclude nongeneric cases like $\alpha_1 = \alpha_3$. We choose

$$\alpha_1 = 0.4, \ \alpha_2 = 1, \ \alpha_3 = 0.6, \ \alpha_4 = 4. \tag{8.56}$$

This choice results in

$$\dot{\chi}_1 = -\frac{1}{8}\varepsilon^2(-0.8r_1^2 + 0.2r_2^2 + (r_2^2 - r_1^2)\cos 2\chi_1). \tag{8.57}$$

Equation (8.57) results in resonance zones generated by χ_1, i.e. small neighbourhoods of

$$M_1 : 3r_1^2 - 2r_2^2 = 0, \ (2\chi_1 = 0, 2\pi), \ M_2 : r_1^2 - 4r_2^2 = 0, \ (2\chi_1 = \pi, 3\pi). \tag{8.58}$$

The combination angles χ_2, χ_3 from Eq. (8.55) are not timelike in a subset of M_1 if we can find values of the amplitudes (or actions) such that $\dot{\chi}_2 = 0$ or $\dot{\chi}_3 = 0$.

The recurrency properties of the phase-flow are strongly dependent on the initial conditions. In Fig. 8.13 we show the Euclidean distance to the initial conditions in two cases. Left we start away from stable periodic solutions (outside M_1), the orbits

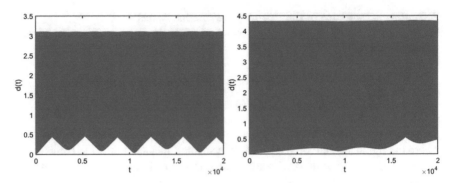

Fig. 8.13 The 1:1:4 resonance in 20,000 timesteps, initial velocities zero, $b_1 = 1, b_2 = -1.5, \varepsilon = 0.1$. Left the Euclidean distance $d(t)$ of the orbits to their initial conditions in the case of initial conditions outside the resonance zones: $q_1(0) = 0.3, q_2(0) = 0.5, q_3(0) = 0.7$. Recurrence takes many timesteps because of passage through the resonance zones. Right the Euclidean distance $d(t)$ for the same Hamiltonian (8.50) but inside primary resonance zone M_1, starting at a location of higher order resonances $q_1(0) = 0.3674, q_2(0) = 0.45, q_3(0) = 1.0129$ ($\cos 2\chi_1 = 1$). First, the recurrence is stronger until the orbit drifts outside the higher order resonance location in M_1. I_3 (and r_3) experiences 0.01 variation

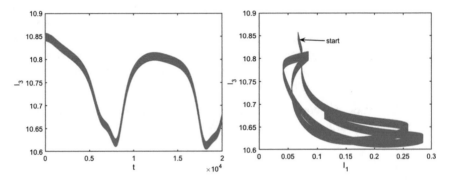

Fig. 8.14 The 1:1:4 resonance in 20,000 timesteps, initial velocities zero, $b_1 = 1, b_2 = -1.5, \varepsilon = 0.1$. Dynamics in the stable part of the primary resonance zone M_1, initial values $q_1(0) = 0.3674, q_2(0) = 0.45, q_3(0) = 1.65$ and initial velocities zero. Left the timeseries $I_3(t)$, right the diagram I_1, I_3 dependent on t. The fluctuations are $O(\varepsilon)$ although the calculations go far beyond $1/\varepsilon^2$

have a recurrence delayed by quasi-trapping in M_1 (see [84]). Right we start with the same Hamiltonian in the resonance zone M_1 but near a location where the higher order combination angles are not timelike; the recurrence is stronger but then the variation of ϕ_3 produces a drift-off of the solutions in M_1 and worse recurrence.

To illustrate the stable and unstable behaviour of the solutions starting in M_1 we present in Fig. 8.14 left the case where $q_3(0) = 0.3$ (χ_2, χ_3 are timelike) and right the unstable case $q_3(0) = 1.0129$. The scales have been adjusted to the variations of $I_1(t), I_3(t)$.

Chapter 9
Quasi-Periodic Solutions and Tori

When studying dynamical systems, defined by differential equations or maps, a basic approach is to locate and to characterise the classical ingredients of such systems. These ingredients are critical points (equilibrium solutions), periodic solutions, invariant manifolds (in particular tori containing quasi-periodic solutions), homoclinics, heteroclinics and in general stable and unstable manifolds emanating from special solutions. As the theory is complex and still in development we will give a more open-ended discussion than in other chapters of our toolbox.

Also we will extend the contraction method, used for periodic solutions by the Poincaré-Lindstedt method, to quasi-periodic solutions and tori. This is a new part of the theory.

Consider a perturbation problem like

$$\dot{x} = f(x) + \varepsilon R(t, x, \varepsilon) \tag{9.1}$$

where as usual ε will indicate a small, positive parameter, R represents a smooth perturbation. Suppose for instance that we have found an isolated torus T_0 if $\varepsilon = 0$ in Eq. (9.1). Does this manifold persist, slightly deformed as a torus T_ε, of Eq. (9.1)?

Or suppose that we have transformed problem (9.1) by introducing variational equations and located a torus for $\varepsilon > 0$ in a corresponding averaged system. Note that the variational equations can also be seen as a perturbation of an averaged or normalised equation. The question can then again be rephrased as the question of persistence of the torus under perturbations.

If the torus T_0 or the torus in the averaged equation is *normally hyperbolic*, the answer is affirmative (normally hyperbolic means loosely speaking that the strength of the flow along the manifold is *weaker* in size than the rates of attraction or repulsion to the manifold). We will discuss such cases. In many applications however the normal hyperbolicity is far from easy to establish.

In the case of Hamiltonian systems the situation is different as the tori arise in families, they are not hyperbolic. Tori for Hamiltonian problems are a special topic

F. Verhulst, *A Toolbox of Averaging Theorems*, Surveys and Tutorials in the Applied Mathematical Sciences 12, https://doi.org/10.1007/978-3-031-34515-9_9

that will not be systematically discussed here, see the extensive surveys [16] and [18].

We will look at different scenarios for the emergence of tori in a number of examples. A torus is often generated by various independent rotational motions - at least two - and we shall find different timescales and quasi- or almost-periodic solutions characterising these rotations.

In addition we stress that invariant manifolds like tori can also show bifurcations; this invokes much more complicated dynamics than bifurcations of equilibria or periodic solutions.

Deforming a Normally Hyperbolic Manifold

Consider the dynamical system in $\mathbb{R}^n$, described by the equation:

$$\dot{x} = f(x)$$

and assume that the system contains a smooth (C^r) invariant manifold M with as important example a torus. The smoothness enables us to define a *tangent* bundle $T(M)$ describing the flow along M and a *normal* bundle $N(M)$ of M, describing the perpendicular flow to M. A typical situation in mechanics involves N coupled two-dimensional oscillators containing a m-dimensional torus with $2 \leq m \leq N$ (the system is described by N amplitudes and N angles). In this case the phasespace dimension is $n = 2N$, the tangent bundle is m-dimensional, the normal bundle $(2N - m)$-dimensional.

Hyperbolicity is introduced as follows. Assume that we can split the corresponding normal bundle of M with respect to the flow generated by the dynamical system completely in an exponentially stable one $N(M)^s$ and an exponentially unstable one $N(M)^u$, with no other components.

In this case the manifold M is called hyperbolic. If this hyperbolic splitting does not contain an unstable manifold $N(M)^u$, M is stable. For a more detailed discussion of these classical matters see for instance Hirsch et al. [37].

Note that the smoothness of M is needed in this description. In many cases the manifolds under consideration loose smoothness at certain bifurcation points when varying parameters. In such cases Lyapunov exponents can still be used to characterise the stability.

The manifold M is moreover *normally hyperbolic* if, measured in the matrix- and vector norms in $\mathbb{R}^n$, $N(M)^u$ expands more sharply than the flow associated with $T(M)$ and $N(M)^s$ contracts more sharply than $T(M)$ under the flow.

Often in the literature the concept of normal hyperbolicity is used without explicit definition or even mentioning the term, but it will be implicitly present in the conditions.

In a number of applications, the situation is simple because a small parameter is present which induces slow and fast dynamics in the dynamical system. Consider the system:

$$\begin{cases} \dot{x} & = \varepsilon f(x, y), \quad x \in D \subset \mathbb{R}^n, t \geq 0 \\ \dot{y} & = g(x, y), \quad y \in G \subset \mathbb{R}^m \end{cases} \tag{9.2}$$

with f and g sufficiently smooth vector functions in x, y. Putting $\varepsilon = 0$ we have $x(t) = x(0)$ and from the second equation $\dot{y} = g(x_0, y)$. For $\varepsilon > 0$ and small we expect $x(t)$ to be slowly varying. Different things happen in the region where $g(x, y)$ vanishes or is $O(\varepsilon)$. An isolated root $y(x)$ of the equation $g(x, y) = 0$ corresponding with a compact manifold will be called a slow manifold. Fenichel (1971–1979) has shown that if this root is hyperbolic with x considered a parameter, it corresponds with a nearby hyperbolic invariant manifold of the full system. We can look for periodic and quasi-periodic solutions or tori on a slow manifold. For more explanation and references see [77].

In the analysis, the fact that if this root is hyperbolic, the corresponding manifold is also normally hyperbolic, is inherent in the problem formulation. For the fibers of the slow manifold are ruled by the fast time variable t, whereas the dynamics of the drift along the manifold is ruled by the timelike variable εt.

9.1   Tori by Bogoliubov-Mitropolsky-Hale Continuation

The branching off of tori is more complicated than the emergence of periodic solutions in dynamical system theory. Emergence of tori was considered extensively by Bogoliubov and Mitropolsky [13], using basically continuation of quasi-periodic motion under perturbations. Another survey together with new results can be found in Hale [34]; see the references there. A formulation in the more general context of bifurcation theory can be found in Chow and Hale [19]. We present several theorems from Hale [34] in an adapted form; see also Hale [33].

Theorem 9.1 *Consider the system S (Hale's notation)*

$$\begin{cases} \dot{\theta} & = \omega(t, \theta) + \varepsilon \omega_1(t, \theta, x, y) + \varepsilon^2 \cdots, \\ \dot{x} & = A(\theta)x + \varepsilon A_1(t, \theta, x, y) + \varepsilon^2 \cdots, \\ \dot{y} & = B(\theta)y + \varepsilon B_1(t, \theta, x, y) + \varepsilon^2 \cdots \end{cases} \tag{9.3}$$

with $\theta \in \mathbb{R}^k, x \in \mathbb{R}^n, y \in \mathbb{R}^m$; all vector functions on the righthand side are periodic in the angle θ and time t.

Such a system arises naturally from local perturbations of differential equations in a neighbourhood of an invariant manifold where the 'unperturbed' system

$$\dot{\theta} = \omega(t, \theta), \ \dot{x} = A(\theta)x, \ \dot{y} = B(\theta)y$$

is assumed to have an invariant manifold M_0 given by

$$M_0 = \{(t, \theta, x, y) : x = y = 0\}.$$

We also assume for system S that

1. *All vector functions on the righthand side of system (9.3) are continuous and bounded; the $\varepsilon^2 \cdots$ terms represent vector functions which are smooth on the domain and which can be estimated $O(\varepsilon^2)$.*
2. *The functions on the righthand side are Lipschitz-continuous with respect to θ, the function $\omega(t, \theta)$ with Lipschitz constant L.*
3. *The functions A_1, B_1, ω_1 are Lipschitz-continuous with respect to x, y.*
4. *There exist positive constants K and α such that for any continuous $\theta(t)$ the fundamental matrices of $\dot{x} = A(\theta)x, \ \dot{y} = B(\theta)y$ can be estimated by $Ke^{-\alpha t}$, $Ke^{\alpha t}$ respectively.*
5. *$\alpha > L$ (normal hyperbolicity).*

Then there exists an invariant manifold M of system S near M_0 with Lipschitz-continuous parametrisation which is periodic in θ.

Note that although α and L are independent of ε their difference may be small. In the applications one should take care that $\varepsilon = o(\alpha - L)$.

Another remark is that Hale's original results are more general than formulated for Theorem 9.1. For instance, the vector functions need not be periodic in θ, but only bounded. If the vector functions are almost-periodic, the parametrisation of M inherits almost-periodicity.

Even more importantly, the perturbations εA_1, εB_1 in the equations for x and y can be replaced by $O(1)$ vector functions. However, this complicates the conditions of the corresponding theorem. Also, to check the conditions in these more general cases is not so easy.

We turn now to a case arising often in applications.

9.2   The Case of Parallel Flow

In a number of important applications the frequency vector $\omega(t, \theta)$ of system S is constant; this will cause the flow on M_0 to be *parallel*. In this case $L = 0$ and the fifth condition of Theorem 9.1 is automatically satisfied.

In addition the case of parallel flow makes it easier to consider cases where the attraction or expansion is weak:

Theorem 9.2 *Consider the system* S_w

$$
\begin{cases}
\dot{\theta} & = \omega + \varepsilon \omega_1(t, \theta, x, y) + \varepsilon^2 \cdots , \\
\dot{x} & = \varepsilon A(\theta)x + \varepsilon A_1(t, \theta, x, y) + \varepsilon^2 \cdots , \\
\dot{y} & = \varepsilon B(\theta)y + \varepsilon B_1(t, \theta, x, y) + \varepsilon^2 \cdots
\end{cases}
\tag{9.4}
$$

with constant frequency vector ω. *As before, this* t- *and* θ-*periodic system is obtained by local perturbation of an invariant manifold* M_0 *in the system*

$$\dot{\theta} = \omega, \quad \dot{x} = \varepsilon A(\theta)x, \quad \dot{y} = \varepsilon B(\theta)y$$

for $x = y = 0$. *In the equations for* x *and* y, $A(\theta)x$ *and* $B(\theta)y$ *represent the linearizations near* $(x, y) = (0, 0)$ *so* A_1, B_1 *are* $o(\|x\|, \|y\|)$. *Assume that*

1. *All vector functions on the righthand side of system* (9.4) *are continuous and bounded; the* $\varepsilon^2 \cdots$ *terms represent vector functions which are smooth on the domain and which can be estimated* $O(\varepsilon^2)$.
2. *The functions on the righthand side are Lipschitz-continuous with respect to* θ, *the function* ω_1 *with Lipschitz constant* η.
3. *The functions* ω_1, A_1, B_1 *are Lipschitz-continuous with respect to* x, y.
4. *There exist positive constants* K *and* α *such that for any continuous* $\theta(t)$ *the fundamental matrices of* $\dot{x} = \varepsilon A(\theta)x, \dot{y} = \varepsilon B(\theta)y$ *can be estimated by* $K e^{-\varepsilon \alpha t}, K e^{\varepsilon \alpha t}$ *respectively.*
5. $\alpha > \eta$ *(normal hyperbolicity at higher order)*

then there exists an invariant manifold M *of system* S_w *near* M_0 *with Lipschitz-continuous parametrisation which is periodic in* θ.

The frequency vector being constant in system S_w enables us to introduce slowly varying phases by putting

$$\theta(t) = \omega t + \psi(t).$$

The resulting system S_w is of the form

$$\dot{X} = \varepsilon F(t, x) + \varepsilon^2 \cdots ,$$

where we have replaced (ψ, x, y) by X. The system is quasi-periodic in t. The near-identity transformation

$$X(t) = z(t) + \varepsilon u(t, z(t)), \quad u(t, z(t)) = \int_0^t (F(t, z(t)) - F_0(z(t)))dt$$

with $F_0(z(t))$ the average over the periods of F in t leads to the equation

$$\dot{z} = \varepsilon F_0(z) + \varepsilon^2 \cdots .$$

Note that as yet we have not introduced any approximation. Usually we can relate Theorem 9.2 to the equation for z which will in general - at least to $O(\varepsilon)$ - be much simpler than the system S_w.

9.3   Tori Created by Neimark-Sacker Bifurcation

An important and practical scenario to create a torus, arises from Neimark-Sacker bifurcation. As a periodic solution may be produced from a bifurcation of an equilibrium with 2 double imaginary eigenvalues, in a similar way a periodic solution with 2 double imaginary eigenvalues may produce a torus. This is the case of a Neimark-Sacker bifurcation, also called Hopf-Hopf bifurcation. For an instructive and detailed introduction see [46]). Suppose that we have obtained an averaged equation $\dot{x} = \varepsilon f(x, a)$, with dimension 3 or higher, by variation of constants and subsequent averaging; a is a parameter or a set of parameters. It is well-known, see Chap. 3, that if this equation contains a critical point, the original equation may contain a periodic solution. The first order approximation of this periodic solution is characterised by the timelike variables t and εt.

Suppose now that by varying the parameter a a pair of eigenvalues of the critical point becomes purely imaginary for $a = a_0$. The variation of a through a_0 should be transversal. For this value of a_0 the averaged equation undergoes a Hopf bifurcation producing a periodic solution of the averaged equation; the typical timelike variable of this periodic solution is εt and so the period will be $O(1/\varepsilon)$. As it branches off an existing periodic solution in the original equation, it will produce a torus; it is associated with a Hopf bifurcation of the corresponding Poincaré map and the bifurcation has a different name: Neimark-Sacker bifurcation. The result will be a two-dimensional torus which contains two-frequency oscillations, one on a timescale of order 1 and the other on timescale $O(1/\varepsilon)$.

9.4   Applications

We start with relatively simple examples.

9.4.1   A Forced Van der Pol-Oscillator

There are natural extensions to non-autonomous systems by introducing the so-called stroboscopic map. We demonstrate this by an example derived from Broer et al. [17]. See also for more background the monograph by Broer et al. [16]. The following example can be extended to other cases.

Consider the forced Van der Pol-oscillator

$$\ddot{x} + x = \mu(1 - x^2)\dot{x} + \varepsilon \cos \omega t$$

which we write as the system

$$\begin{cases} \dot{x} &= y, \\ \dot{y} &= -y + \mu(1 - x^2)y + \varepsilon \cos \tau, \\ \dot{\tau} &= \omega. \end{cases} \tag{9.5}$$

The 2π-periodic forcing term $\varepsilon \cos \tau$ produces a stroboscopic map of the x, y-plane into itself. For $\varepsilon = 0$ this is just the map of the periodic solution of the Van der Pol-equation, an invariant circle, into itself and the closed orbit is normally hyperbolic. In the extended phase space $\mathbb{R}^2 \times \mathbb{R}/2\pi\mathbb{Z}$ this invariant circle for $\varepsilon = 0$ corresponds with a normally hyperbolic torus of system (9.5) which is persistent for small, positive values of ε.

Actually, the authors, choosing $\mu = 0.4, \omega = 0.9$ consider what happens if ε increases. At $\varepsilon = 0.3634$ the normal hyperbolicity is destroyed by a saddle-node bifurcation.

Extension to Coupled Van der Pol-Oscillators
An example of a normally hyperbolic torus in a coupled system is found in the system:

$$\begin{cases} \ddot{x} + x &= \mu(1 - x^2)\dot{x} + \varepsilon f(x, y), \\ \ddot{y} + \omega^2 y &= \mu(1 - y^2)\dot{y} + \varepsilon g(x, y), \end{cases} \tag{9.6}$$

with ε-independent positive constant ω and μ (fixed positive numbers, $O(1)$ with respect to ε) and smooth perturbations f, g. Omitting the perturbations f, g we have two uncoupled normally hyperbolic oscillations. In general, if ω is irrational, the combined oscillations attract to a torus in 4-space, the product of the two periodic attractors, filled with quasiperiodic motion. Adding the perturbations f, g can not destroy this two-dimensional torus but only deforms it.

If μ is large enough, the tori are for $\varepsilon = 0$ built up from relaxation oscillations. The timescales of rotation are also different in this case, see [30].

9.4.2 Quasi-Periodic Solutions in the Forced Van der Pol-Equation

Consider the equation:

$$\ddot{x} + x = \varepsilon(1 - x^2)\dot{x} + a \cos \omega t$$

with a and ω constants. This time we assume that the nonlinearity is small and the forcing can be $O(1)$ as $\varepsilon \to 0$.

Assume ω is not ε-close to 1, for if ω is near to 1, the solutions move away from an $O(1)$ neighbourhood of the origin because of linear resonance. We introduce the standard transformation $x, \dot{x} \to r, \psi$ as in (1.27):

$$x = r\cos(t + \phi) + \frac{a}{1 - \omega^2}\cos\omega t, \ \dot{x} = -r\sin(t + \phi) - \frac{a\omega}{1 - \omega^2}\sin\omega t.$$

The resulting slowly varying system can be averaged, producing periodic solutions in which various values of ω play a part. Returning to the corresponding expressions for x and $\dot{x}$ we conclude to the presence of tori in the phase space extended with t.

The approximate behaviour is studied in Sect. 5.3.4. Averaging produces in this case:

$$\dot{r} = \frac{\varepsilon}{2}r(1 - \frac{1}{4}r^2 - \frac{1}{2}\frac{a^2}{(1 - \omega^2)^2}) + \varepsilon^2 \cdots, \ \dot{\psi} = \varepsilon^2 \cdots, \tag{9.7}$$

Choosing as an example $a = 0.4, \omega = \sqrt{2}$ we have for the approximate r:

$$\dot{r} = (0.92 - \frac{1}{4}r^2) + \varepsilon^2 \cdots$$

The amplitude r tends to 1.918. In the case of $a = O(1)$ the corresponding solutions will be ε-close to the torus described in this case by

$$x = 1.918\cos(t + \phi(0)) - \frac{2}{5}\cos\sqrt{2}t, \ \dot{x} = -1.918\sin(t + \phi(0)) + \frac{2\sqrt{2}}{5}\sin\sqrt{2}t.$$

9.4.3 Neimark-Sacker Bifurcation in Two Coupled Oscillators

The analysis becomes more complicated if we have not already a torus structure for $\varepsilon = 0$ that we can perturb to start with. We will describe a few results for autonomous coupled oscillators of the paper [7]:

$$\begin{cases} \ddot{x} + x + \varepsilon(axy + \delta x^2\dot{x} + \gamma x^3) = 0, \\ \ddot{y} + 4y + \varepsilon(\kappa\dot{y} + bx^2) = 0. \end{cases} \tag{9.8}$$

We can rescale x to obtain $a = 1$. The x-oscillator has a nonlinear damping, the y-oscillator a linear one ($\kappa, \delta \geq 0$), the coupling is nonlinear. The strategy is to find a periodic solution by averaging; to determine its stability is possible analytically but easier using numerical continuation, for instance by MATCONT [47], while determining the Floquet (Lyapunov) exponents at each step. When the exponents correspond with a Neinark-Sacker bifurcation we expect the emergence of a 2-torus. A bonus in this case is that continuation also reveals destruction of tori and the emergence of a strange attractor. It will turn out that the parameter γ is suitable for continuation, probably because the term γx^3 changes the oscillation period of the x-oscillator in a nonlinear way.

Transforming to amplitude-phase coordinates (1.6) $x = r_1 \cos(t + \phi_1)$ etc. and averaging we find with combination angle $\chi = 2\phi_1 - \phi_2$:

$$\begin{cases} \dot{r}_1 = \varepsilon r_1(\frac{1}{4}r_2 \sin \chi - \frac{\delta}{8}r_1^2), \\ \dot{r}_2 = \frac{\varepsilon}{2}r_2(-b\frac{r_1^2}{4r_2} \sin \chi - \kappa), \\ \dot{\chi} = \varepsilon(\frac{1}{2}r_2 \cos \chi + \frac{3}{4}\gamma r_1^2 - b\frac{r_1^2}{8r_2} \cos \chi). \end{cases} \tag{9.9}$$

Putting the righthand sides of system (9.9) equal to zero we look for solutions outside the coordinate planes. From the first 2 equations we find nontrivial solutions if $b < -2\kappa\delta < 0$, b has to be negative; the third equation in (9.9) implies $\cos \chi < 0$. We find for $\kappa > 0$ necessary conditions for equilibria:

$$r_2 = -\frac{b}{4\kappa}r_1^2 \sin \chi, \; r_1 = \sqrt{\frac{-8\kappa^2 \cot \chi}{12\kappa\gamma - b \sin 2\chi}}. \tag{9.10}$$

It is clear that we have the condition $\sin \chi \geq 0$. If $\kappa = 0$ we find a continous family of equilibria and at this order of approximation not an isolated periodic solution. System (9.8) has an isolated periodic solution if $\kappa, \delta > 0$, $b < -2\kappa\delta$ and $\pi/2 < \chi < \pi.$.

A surprising result in [7] obtained by using the Routh-Hurwitz criterion is that this isolated periodic solution is asymptotically stable. This result is valid for small ε and would *exclude* a torus bifurcation, so it might produce new results when considering larger values of ε. At this point the combination of asymptotic analysis and numerical bifurcation methods, in our case MATCONT [47], becomes very fruitful. As an example we choose in system (9.8) $\varepsilon = 1, a = 0.5, b = -0.5, \delta = 0.4, \kappa = 0.1$. We use γ as bifurcation parameter, starting with $\gamma = 0.2$ and slowly decreasing its value. Near the value $\gamma = 0.096$ we detect a Neimark-Sacker bifurcation producing a stable torus that is not very close to the origin of phase-space. Decreasing γ we find at smaller values, $\gamma = 0.0892$, torus destruction while a strange attractor emerges, see Fig. 9.1. In [7] there are more pictures, in particular Poincaré sections, and more technical details like period doublings and Arnold tongues are discussed.

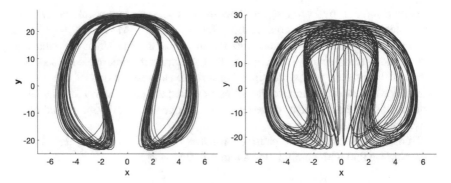

Fig. 9.1 Two projections from 4-space on the x, y-plane of system (9.8). We took $a = 0.5$, $b = -0.5$, $\delta = 0.4$, $\kappa = 0.1$ and in both figures $x(0) = 1$, $\dot{x}(0) = 0$, $y(0) = 22$, $\dot{y}(0) = 0$. If $\gamma > 0.096$ we find a stable periodic solution, but at $\gamma = 0.096$ a Neimark-Sacker bifurcation produces a torus (left). If $\gamma = 0.0892$ we find a strange attractor (right) with Kaplan-Yorke dimension $D_{KY} = 2.32$. Note that the initial conditions produce some transient orbits

9.4.4   Interaction of Vibrations and Parametric Excitation

A simplified case of a system involving parametric excitation studied in [5] is the system:

$$\begin{cases} \ddot{x} + \varepsilon \kappa_1 \dot{x} + (1 + \varepsilon \cos 2t)x + \varepsilon(axy + \gamma x^3) & = 0, \\ \ddot{y} + \varepsilon \kappa_2 \dot{y} + 4(1 + \varepsilon)y - \varepsilon b x^2 & = 0. \end{cases} \tag{9.11}$$

This is a system with parametric excitation and nonlinear coupling; $\kappa_1, \kappa_2 \geq 0$ damping coefficients independent of ε. The oscillators are containing the $1 : 2$ resonance but with small detuning.

If $\kappa_1, \kappa_2 > 0$ the origin is stable, the $(y, \dot{y})$ coordinate plane contains a damped, linear oscillator. Away from the coordinate planes we may use the usual amplitude-phase variables by transformation (1.6). First order averaging produces the approximating system:

$$\begin{cases} \dot{r}_1 & = \varepsilon r_1 \left(a \frac{r_2}{4} \sin(2\phi_1 - \phi_2) + \frac{1}{4} \sin 2\phi_1 - \frac{1}{2}\kappa_1 \right), \\ \dot{\phi}_1 & = \varepsilon \left(a \frac{r_2}{4} \cos(2\phi_1 - \phi_2) + \frac{1}{4} \cos 2\phi_1 + \frac{3}{8}\gamma r_1^2 \right), \\ \dot{r}_2 & = \varepsilon \frac{r_2}{2} \left(b \frac{r_1^2}{4r_2} \sin(2\phi_1 - \phi_2) - \kappa_2 \right), \\ \dot{\phi}_2 & = \frac{\varepsilon}{2} \left(-b \frac{r_1^2}{4r_2} \cos(2\phi_1 - \phi_2) + 2 \right). \end{cases} \tag{9.12}$$

To study periodic solutions near the coordinate planes (normal modes) we have to use a different coordinate transformation, but to locate a torus we are more interested in general position periodic orbits. In the averaged system we obtain the combination angle $\chi = 2\phi_1 - \phi_2$.

Putting the righthand sides equal to zero produces various solutions, but as we have 5 free parameters the linear algebra becomes already complicated. A systematic study in [5] produces stable and unstable periodic solutions.

A simple but rather degenerate case is given bij the choice:

$$a = b = 1, \gamma = -\frac{1}{6}, \kappa_1 = \kappa_2 = 0, \sin \chi = \sin 2\phi_1$$

$$= 0, \cos \chi = \cos 2\phi_1 = 1, r_1 = 2\sqrt{2}, r_2 = 1.$$

Linearising near this critical point of the vector field of system (9.12) we find the eigenvalues:

$$\lambda_{1,2} = \lambda_{3,4} = \pm 0.7071i,$$

This symmetric result is caused by the absence of damping, the flow in the extended phase-space is measure-preserving.

Consider the periodic solution obtained by the choice: $a = 1, b = 4, \kappa_1 = \gamma = 0, \kappa_2 = 1$ leading to the crtical point given by:

$$\sin \chi = -\sin 2\phi_1 = \frac{1}{\sqrt{5}}, \cos \chi = -\cos 2\phi_1 = \frac{2}{\sqrt{5}}, r_1^2 = \sqrt{5}, r_2 = 1.$$

Linearising near this critical point of the vector field of system (9.12) we find the eigenvalues:

$$\lambda_{1,2} = 0.09444 \pm 0.9384i, \lambda_3 = -0.0354, \lambda_4 = 0.5615.,$$

These special cases are instructive but rather laborious so it is surprising that in [5] the averaging result of system (9.12) can be used to find explicitly cases of periodic solutions with 2 purely imaginary eigenvalues. Systematically using MATCONT [47] and increasing the coefficients κ_1, κ_2 from small positive values we find that for critical values of the parameters the real part of two eigenvalues vanishes. The critical point itself corresponds with a periodic solution but the critical value corresponds with a Hopf bifurcation of the periodic solution . This in its turn corresponds with a torus in the original equations (in x and y) which is a Neimark-Sacker bifurcation. For more details see [5].

9.4.5   *Interaction of Self-Excited and Parametric Excitation*

In a long series of books (partly published by the Bechovice Research Institute) and articles, A. Tondl, a Czech engineer, constructed models describing nonlinear phenomena in engineering. One of his many topics was concerned with the interaction of self-excited and parametric vibrations, see [62]. In a private communication A. Tondl gave in 2002 a model for undesirable flow-induced vibrations ($y_2(t)$) that might be influenced or reduced by coupling to parametric vibration absorbers y_1, y_3. The system for the deflections y_1, y_2, y_3 is:

$$\begin{cases} m_1 \ddot{y}_1 + b_0 \dot{y}_1 + k_0(1 + \varepsilon \cos \omega t)y_1 - k_1(y_2 - y_1) & = 0, \\ m_2 \ddot{y}_2 + \beta U^2(1 - \gamma \dot{y}_2^2)\dot{y}_2 + 2k_1 y_2 - k_1(y_1 + y_3) & = 0, \\ m_3 \ddot{y}_3 + b_0 \dot{y}_3 + k_0(1 + \varepsilon \cos \omega t)y_3 - k_1(y_2 - y_3) & = 0. \end{cases} \quad (9.13)$$

All parameters are assumed to be positive, the flow-induced vibrations of mass m_2 with flow strength U are caused by a so-called Raleigh self-excitation term. The idea is to quench the flow-induced vibrations of mass m_2 by nonlinear coupling to 2 parametrically excited oscillators with masses m_1, m_3. The analysis is complicated, uses MATHEMATICA and MATCONT leading to periodic solutions that by Neimark-Sacker bifurcation evolve to tori; slowly changing the parameters in this 3-mass system one finds break-up of the tori and evolution to strange attractors. This is basically an instructive example of the Ruelle-Takens scenario [55] that sketches the origin of turbulence by such a sequence of bifurcations from unstable periodic solutions to tori to strange attractors. For extensive details of this flow-induced vibration model see [8]. Omitted in this study are details of the case $m_1 = m_3$ but it is shown in [8] that the averaged system of (9.13) decouples and can be treated as a 2-dof system.

A Cartoon Problem of Self-Excited and Parametric Excitation Interaction
Instead of giving more details of the complicated system (9.13) we consider a highly simplified system that contains partly the same elements, it is discussed in [10]. Consider the system:

$$\begin{cases} \ddot{x} + \varepsilon \kappa \dot{x} + (1 + \varepsilon \cos 2t)x + \varepsilon c y & = 0, \\ \ddot{y} - \varepsilon \mu(1 - y^2)\dot{y} + y + \varepsilon c x & = 0. \end{cases} \quad (9.14)$$

We have weak linear interaction between the 2 oscillators governed by parameter c, with parameters $\kappa, \mu \geq 0$. General position periodic solutions are candidates to produce tori. Using transformation (1.6) we find variational equations for amplitudes r_1, r_2 and phases ϕ_1, ϕ_2. We leave out the variational equations. After averaging over time we find:

$$\begin{cases} \dot{r}_1 &= \frac{\varepsilon}{2}(-\kappa r_1 + \frac{1}{2}r_1 \sin 2\phi_1 + cr_2 \sin \chi), \\ \dot{\phi}_1 &= \frac{\varepsilon}{2r_1}(\frac{1}{2}r_1 \cos 2\phi_1 + cr_2 \cos \chi), \\ \dot{r}_2 &= \frac{\varepsilon}{2}(\mu r_2(1 - \frac{1}{4}r_2^2) - cr_1 \sin \chi), \\ \dot{\phi}_2 &= -\frac{\varepsilon}{2}c\frac{r_1}{r_2} \cos \chi, \end{cases} \tag{9.15}$$

with $\chi = \phi_1 - \phi_2$. The averaged system has a general position critical point from the equation for ϕ_2 if $\cos \chi = 0$ or $\sin \chi = \pm 1$ from the equation for ϕ_1. The other conditions for critical points become in this case:

$$-\kappa r_1 + \frac{1}{2}r_1 \sin 2\phi_1 + cr_2 \sin \chi = 0, \quad \mu r_2(1 - \frac{1}{4}r_2^2) - cr_1 \sin \chi = 0, \quad \cos 2\phi_1 = 0. \tag{9.16}$$

We have, for the critical values:

$$r_1 = \frac{-c \sin \chi}{-\kappa + 0.5 \sin 2\psi_1} r_2, \; r_2 = 2\sqrt{1 + \frac{c^2}{\mu(-\kappa + 0.5 \sin 2\psi_1)}}, \; \sin 2\psi_1 = \pm 1. \tag{9.17}$$

Nontrivial equilibria can be found for $\sin \chi = 1$, $c > 0$, $\sin 2\phi_1 = \pm 1$, $\kappa > 0$ and $\mu > 0$ large enough. Equation (9.17) becomes in the case $\sin 2\psi_1 = -1$:

$$r_1 = \frac{c}{\kappa + 0.5} r_2, \; r_2 = 2\sqrt{1 - \frac{c^2}{\mu(\kappa + 0.5)}}. \tag{9.18}$$

The Jacobian matrix becomes at the nontrivial equilibrium:

$$\begin{pmatrix} -\kappa - \frac{1}{2} & 0 & -c & 0 \\ 0 & \frac{1}{2} - \kappa & 0 & \kappa + \frac{1}{2} \\ c & 0 & \frac{6c^2}{2\kappa-1} - 2\mu & 0 \\ 0 & -\frac{2c^2}{2\kappa+1} & 0 & \frac{2c^2}{2\kappa+1} \end{pmatrix} \tag{9.19}$$

The eigenvalues at the equilibrium can in this case be computed by MATHEMATICA. We find from [10]:

$$\lambda_1 = \frac{-\sqrt{-8c^2(4\kappa(\kappa+2)+3)+16c^4+(1-4\kappa^2)^2} + 4c^2 - 4\kappa^2 + 1}{8\kappa + 4}, \tag{9.20}$$

$$\lambda_2 = \frac{\sqrt{-8c^2(4\kappa(\kappa+2)+3)+16c^4+\left(1-4\kappa^2\right)^2}+4c^2-4\kappa^2+1}{8\kappa+4}, \tag{9.21}$$

$$\lambda_3 = \frac{-\sqrt{8c^2(2\kappa-1)(2\kappa-12\mu+5)+144c^4+(1-2\kappa)^2(2\kappa-4\mu+1)^2}+R}{8\kappa-4}, \tag{9.22}$$

$$\lambda_4 = \frac{\sqrt{8c^2(2\kappa-1)(2\kappa-12\mu+5)+144c^4+(1-2\kappa)^2(2\kappa-4\mu+1)^2}+R}{8\kappa-4} \tag{9.23}$$

$$R = 12c^2 - 4\kappa^2 - 8\kappa\mu + 4\mu + 1. \tag{9.24}$$

From Eqs. (9.20) and (9.21) we conclude that the real part (i.e. the part outside the square root symbol) of λ_1 and λ_2 becomes zero in 2 cases:

$$c(\kappa) = 1/2\sqrt{-1+4\kappa^2}, \quad c(\kappa,\mu) = \frac{1}{2}\sqrt{(1+4\kappa+4\mu+4\kappa^2+8\nu\kappa\mu)/3},$$

with $\kappa \geq \frac{1}{2}$. The term under the square root in the first case is negative for $\kappa > \frac{1}{2}$ so that the parameter curve $c(\kappa) = 1/2\sqrt{-1+4\kappa^2}$ is a Hopf curve in the averaged system and consequently a Neimark-Sacker bifurcation in the original system leading to torus dynamics. Along this curve the eigenvalues λ_1 and λ_2 are purely imaginary. Adding a small perturbation $\delta \ll 1$ to c yields a transversal crossing of the imaginary axis and definitely a Hopf bifurcation. Remarkably this curve is independent of the self-excitation parameter μ. An example and figures can be found in [10]. The use of the expression for $c(\kappa)$ leads to a stable torus where as the use of the expression for $c(\kappa,\mu)$ also leads to a Neimeark-Sacker curve in parameter-space but an unstable torus.

A second equilibrium arises if $\sin\chi = 1$, $\sin 2\phi_1 = 1$. We leave out the calculations of eigenvalues and Neimark-Sacker curve in this case as they also lead to an instable torus, see [10]. For reference we summarise some results in this case:

$$r_1 = \frac{c}{\kappa-0.5}r_2, \quad r_2 = 2\sqrt{1+\frac{c^2}{\mu(-\kappa+0.5)}}. \tag{9.25}$$

For the Neimark-Sacker curve in parameter-space we find:

$$c(\kappa,\mu) = \frac{1}{2}\sqrt{(2\kappa-1)(2\kappa+4\mu-1)/3},$$

9.4.6   Interaction of Self-Excited Oscillations (Hale's Example)

In [34] interaction of self-excited oscillators is considered, for more details see also
[77]. One considers the system:

$$\begin{cases} \ddot{x} + x & = \varepsilon(1 - x^2 - ay^2)\dot{x}, \\ \ddot{y} + \omega^2 y & = \varepsilon(1 - y^2 - bx^2)\dot{y}. \end{cases} \tag{9.26}$$

The normal mode planes are invariant manifolds containing a self-excited Van der
Pol periodic solution. We consider the solutions in general position.

Suppose that $\omega \neq 1$ and not ε-close to 1; $0 < a, b \leq 1$.

We can characterise 2 angles in system (9.26) following the analysis of averaging
over angles from Chap. 7. Transformation (1.21):

$$x = r_1 \sin\phi_1, \dot{x} = r_1 \cos\phi_1, y = r_2 \sin\phi_2, \dot{y} = \omega r_2 \cos\phi_2$$

yields:

$$\begin{cases} \dot{r}_1 & = \varepsilon(1 - r_1^2 \sin^2\phi_1 - ar_2^2 \sin^2\phi_2)r_1 \cos^2\phi_1, \\ \dot{\phi}_1 & = 1 - \frac{\varepsilon}{r_1}(1 - r_1^2 \sin^2\phi_1 - ar_2^2 \sin^2\phi_2)r_1 \sin\phi_1 \cos\phi_1, \\ \dot{r}_2 & = \frac{\varepsilon}{\omega}(1 - r_2^2 \sin^2\phi_2 - br_1^2 \sin^2\phi_1)\omega r_2 \cos^2\phi_2, \\ \dot{\phi}_2 & = \omega - \frac{\varepsilon}{\omega}(1 - r_2^2 \sin^2\phi_2 - br_1^2 \sin^2\phi_1)\omega r_2 \sin\phi_2 \cos\phi_2. \end{cases} \tag{9.27}$$

The vector field contains the angles $\phi_1, \phi_2, \phi_1 - \phi_2$. With our assumptions on ω we
can average over the angles, there is no resonance manifold. We find after averaging:

$$\begin{cases} \dot{r}_1 & = \frac{\varepsilon}{2}r_1(1 - \frac{1}{4}r_1^2 - \frac{1}{2}ar_2^2), \dot{\phi}_1 = 1 + \varepsilon \dots \\ \dot{r}_2 & = \frac{\varepsilon}{2}r_2(1 - \frac{1}{4}r_2^2 - \frac{1}{2}br_1^2), \dot{\phi}_2 = \omega + \varepsilon \dots \end{cases} \tag{9.28}$$

Note that to lowest order the angles depend only on the basic frequencies $1, \omega$. The
righthand sides of system (9.28) vanish to $O(\varepsilon)$ if

$$r_1^2 = \frac{4 - 8a}{1 - 4ab}, r_2^2 = \frac{4 - 8b}{1 - 4ab}, 0 < a, b < \frac{1}{2}. \tag{9.29}$$

These radius values together with the 2 moving angles describe a torus of the
averaged system (9.28). Linearisation of system (9.28) at the radius values (9.29)
produces the characteristic equation:

$$\lambda^2 + \frac{1}{2}(r_1^2 + r_2^2)\lambda + \frac{1}{4}(1 - ab)r_1^2 r_2^2 = 0.$$

We conclude that the torus is normally hyperbolic, persists for the original system (9.1) and is stable.

Remark 9.1 Consider the special case ω not ε-close to 1 but $a = b = \frac{1}{2}$. In this case the torus of system (9.28) collapses to a sphere in phase-space described by $r_1^2 + r_2^2 = 4$. This is a degenerate case as r_1, r_2 restricted to the sphere are still free to choose. Higher order approximations may change the dynamics qualitatively.

As an illustration of the analysis in the case $\omega = 1$ we reproduce a result from [77] ex. 12.13. Using transformation (1.6) to r_1, r_2, ϕ_1, ϕ_2 we find after first-order averaging:

$$
\begin{cases}
\dot{r}_1 = \frac{\varepsilon}{2} r_1 (1 - \frac{1}{4} r_1^2 - \frac{a}{2} r_2^2 + \frac{a}{4} r_2^2 \cos 2\chi), \\
\dot{r}_2 = \frac{\varepsilon}{2} r_2 (1 - \frac{1}{4} r_2^2 - \frac{b}{2} r_1^2 + \frac{b}{4} r_1^2 \cos 2\chi), \\
\dot{\chi} = -\frac{\varepsilon}{4} (b r_1^2 + a r_2^2) \sin 2\chi,
\end{cases}
\tag{9.30}
$$

with $\chi = \phi_1 - \phi_2$. Putting $\sin 2\chi = 0$ we find periodic solutions.

We can also look for exact solutions by putting $y(t) = px(t)$; we find a continuous family of p-values producing exact Van der Pol periodic solutions in system (9.26) with $\omega = 1$. If for instance $a = b = \frac{1}{2}$ we have the family of exact solutions $y(t) = \pm x(t)$ in phase-space that contains attracting periodic solutions.

9.5   Iteration by Integral Equations

We consider again Eq. (1.1) for n-dimensional ODEs:

$$
\dot{x} = f_0(t, x) + \varepsilon f_1(t, x) + O(\varepsilon^2).
$$

As we have seen in the Introduction, simple expansion with respect to ε to obtain approximations of the solutions runs into difficulties. The basic idea of the Poincaré-Lindstedt method, Sect. 3.3, is to use the Poincaré-expansion theorem in combination with iteration via a contraction argument (see Introduction) on a suitable integral equation. The expansions produce approximation results on the timescale 1, so it is essential to focus on solutions that can be characterised on such a short timescale, in Sect. 3.3 we considered periodic solutions. The periodicity conditions are directly related to averaging.

We can use this idea also to approximate quasi-periodic solutions and bounded invariant manifolds of Eq. (1.1), for instance tori. In this case we have to start with a suitable solution of the equation if $\varepsilon = 0$ with n free parameters λ_0; suppose this solution is $F(t, \lambda_0)$. We can formulate the equivalent Volterra integral equations or use variation of constants to take profit of the behaviour of the unperturbed solution $F(t, \lambda_0)$. We have some freedom in formulating the integral equations. In both cases we use iteration and Poincaré-expansion while applying *quasi-periodic secularity conditions* on the parameters λ_0.

We will make the procedure more explicit for N dof systems of the form:

$$\ddot{x}_i + \omega_i^2 x_i = \varepsilon f(t, x, \dot{x}), \tag{9.31}$$

with $i = 1, 2, \ldots, N, x = (x_1, x_2, \ldots, x_N)$. Cases of particular interest are when some pairs of constant, positive frequencies $\omega_i, i = 1, 2, \ldots, N$ have irrational ratio's. For such a $2N$ dimensional system we use the variational equations from the Introduction. To be more explicit, suppose amplitude-phase transformations (1.6) are suitable. Expanding the parameters r_i, ϕ_i in system (9.31) we have at zero order $x_i^0(t) = r_i^0 \cos(\omega_i t + \phi_i^0)$ and to next order:

$$\tilde{x}(t) = r_i^0 \cos(\omega_i t + \phi_i^0) + \varepsilon \int_0^t \sin(\omega_i t - \omega_i s) f(s, x_i^0(s), \dot{x}_i^0)(s))ds,$$

$$i = 1, 2, \ldots, N. \tag{9.32}$$

If system (9.32) contains external forcing we can adapt the integral equation using appropriate variational transformations, see Example 9.1 below. For the solutions x_i we will apply *quasi-periodic secularity conditions* that will produce at first order $2N$ equations with $2N$ parameter values r_i^0, ϕ_i^0. Solving the $2N$ (in general) transcendental equations and computing the Jacobian of the system we may find that its rank is $2N$. In this case the implicit function theorem applies and we have pinpointed a quasi-periodic solution of system (9.31). The approximation of this solution is $\tilde{x}_i(t)$ with explicit parameter values substituted.

If the rank of the Jacobian matrix is $2N-1$ we have to this order of approximation a branching of the solutions and in some cases a family of quasi-periodic solutions on a bounded manifold. This will often be a torus.

Remark 9.2   If some of the equations of system (9.31) are autonomous we expect a time-dependent shift of the phase. However, this is a complication that arises at $O(\varepsilon)$ and so for $O(\varepsilon^2)$ in the corresponding integral equation. We met this behaviour already in Sect. 3.3 for the Van der Pol-equation where we have the approximation for the phase $\phi(t) = \phi_0 + O(\varepsilon)$ with for the periodic solution perturbation terms dependent on t.

As a 1-dimensional example consider again a case of the forced Van der Pol-equation of Sect. 5.3.4.

Example 9.1

$$\ddot{x} + x = \varepsilon(1 - x^2)\dot{x} + a \cos \sqrt{2}t. \tag{9.33}$$

If $\varepsilon = 0$ we have the solution

$$x^0(t) = r_0 \cos(t + \phi_0) - a \cos \sqrt{2}t,$$

with constants r_0, ϕ_0. Consider this 'unperturbed' solution as the lowest order approximation of the iteration scheme for the integral equation:

$$x(t) = r \cos(t + \phi) - a \cos \sqrt{2}t + \varepsilon \int_0^t \sin(t - s)(1 - x^2(s))\dot{x}(s)ds, \qquad (9.34)$$

with Poincaré-expansions $r = r_0 + \varepsilon r_1 + O(\varepsilon^2)$, $\phi = \phi_0 + \varepsilon \phi_1 + O(\varepsilon^2)$. The second order approximation is obtained by replacing x by the lowest order approximation:

$$\tilde{x}(t) = r_0 \cos(t + \phi_0) - a \cos \sqrt{2}t + \varepsilon I(t), \qquad (9.35)$$

with

$$I(t) = \int_0^t \sin(t - s)(1 - (r_0 \cos(s + \phi_0) - a \cos \sqrt{2}s)^2)$$
$$\times (-r_0 \sin(s + \phi_0) + a\sqrt{2} \sin \sqrt{2}s)ds.$$

Expanding $\sin(t - s) = \sin t \cos s - \cos t \sin s$ the integral splits into 2 parts. To find and approximate quasi-periodic solutions we have to choose r_0, ϕ_0 so that the integrals produce no other type of solution. These are the *quasi-periodic secularity conditions*. In other words, we have to choose the constants such that the generalised Fourier expansion of the integrands contain oscillating terms only. Evaluating

$$\int_0^t \cos s (1 - (r_0 \cos(s + \phi_0) - a \cos \sqrt{2}s)^2)(-r_0 \sin(s + \phi_0) + a\sqrt{2} \sin \sqrt{2}s)ds,$$

we find the condition:

$$r_0 \sin \phi_0 (1 - \frac{1}{2}a^2 - \frac{1}{4}r_0^2) = 0.$$

Evaluating

$$\int_0^t \sin s (1 - (r_0 \cos(s + \phi_0) - a \cos \sqrt{2}s)^2)(-r_0 \sin(s + \phi_0) + a\sqrt{2} \sin \sqrt{2}s)ds,$$

we find the condition:

$$r_0 \cos \phi_0 (1 - \frac{1}{2}a^2 - \frac{1}{4}r_0^2) = 0.$$

We conclude that if $0 < a^2 < 2$ we have an $O(\varepsilon)$ approximate family of quasi-periodic solutions of the form:

$$\tilde{x}(t) = 2\sqrt{1 - \frac{1}{2}a^2} \, \cos(t + \phi_0) - a\sqrt{2}t. \tag{9.36}$$

If $a^2 > 2$ we have a periodic solution of the form $-a \cos \sqrt{2}t$. The 'standard' Van der Pol periodic solution has been destroyed by quasi-periodic forcing.

9.6 Applications of the Iteration Procedure

We will analyse some examples of the form (9.31) with $N = 2$.

9.6.1 Hale's Example by Iteration

Consider iteration for integral equations in the case of Hale's Example 9.4.6 for system (9.26):

$$\ddot{x} + x = \varepsilon(1 - x^2 - ay^2)\dot{x},$$
$$\ddot{y} + \omega^2 y = \varepsilon(1 - y^2 - bx^2)\dot{y}.$$

Frequency ω is irrational and not ε-close to 1, $0 < a, b < 1$. The equivalent integral equations are:

$$\begin{cases} x(t) = r\cos(t + \phi) + \varepsilon \int_0^t \sin(t - s)(1 - x^2(s) - ay^2(s))ds, \\ y(t) = R\cos(\omega t + \psi) + \varepsilon \int_0^t \sin(\omega t - \omega s)(1 - y^2(s) - bx^2(s))ds \end{cases} \tag{9.37}$$

We exclude the normal modes and expand $x(t)$, $y(t)$ and the parameters $r = r_0 + \varepsilon r_1 + O(\varepsilon^2)$, $\phi = \phi_0 + \varepsilon\phi_1 + O(\varepsilon^2)$, $R = R_0 + \varepsilon R_1 + O(\varepsilon^2)$, $\psi = \psi_0 + \varepsilon\psi_1 + O(\varepsilon^2)$. At lowest order we find for the x-approximation:

$$\tilde{x}(t) = r_0 \cos(t + \phi_0)$$

$$- \varepsilon \int_0^t \sin(t - s)(1 - r_0^2 \cos^2(s + \phi_0) - aR_0^2 \cos^2(\omega s + \psi_0))r_0 \sin(s + \phi_0)ds.$$

The integral consists of 2 parts:

$$- \varepsilon r_0 \sin t \int_0^t (1 - r_0^2 \cos^2(s + \phi_0) - aR_0^2 \cos^2(\omega s + \psi_0))\frac{1}{2}\sin(2s + 2\phi_0)ds.$$

and

$$\varepsilon r_0 \cos t \int_0^t (1 - r_0^2 \cos^2(s + \phi_0) - a R_0^2 \cos^2(\omega s + \psi_0)) \sin^2(s + \phi_0) ds.$$

The first integral is uniformly bounded and gives no secularity condition, the second integral gives the quasi-periodic secularity condition:

$$\frac{1}{2} r_0^2 + a R_0^2 = 4. \tag{9.38}$$

At lowest order we have the y-approximation:

$$\tilde{y}(t) = R_0 \cos(\omega t + \psi_0) - \varepsilon \int_0^t \sin(\omega t - \omega s)(1 - R_0^2 \cos^2(\omega s + \psi_0)$$

$$- b r_0^2 \cos^2(s + \phi_0)) R_0 \sin(\omega s + \psi_0) ds.$$

The integral has 2 parts:

$$- \varepsilon R_0 \sin \omega t \int_0^t (1 - R_0^2 \cos^2(\omega s + \psi_0) - b r_0^2 \cos^2(s + \phi_0)) \frac{1}{2} \sin(2\omega s + 2\psi_0) ds.$$

and

$$\varepsilon R_0 \cos \omega t \int_0^t (1 - R_0^2 \cos^2(\omega s + \psi_0) - b r_0^2 \cos^2(s + \phi_0)) \sin^2(\omega s + \psi_0) ds.$$

The first integral is uniformly bounded and gives no condition, the other integral gives the second quasi-periodic secularity condition:

$$\frac{1}{2} R_0^2 + b r_0^2 = 4. \tag{9.39}$$

In r_0, R_0 parameter-space the 2 secularity conditions produce 2 ellipses. Only positive values of r_0, R_0 are useful, we find one solution if $0 < a < \frac{1}{2}, 0 < b < \frac{1}{2}$ and if $\frac{1}{2} < a < 1, \frac{1}{2} < b < 1$; these values correspond with 2-dimensional tori in phase-space. If $a = b = \frac{1}{2}$ we find an invariant sphere.

The first order approximations for $x(t), y(t)$ can be obtained by performing the integrations of $\tilde{x}(t), \tilde{y}(t)$ using the quasi-periodic secularity conditions. The results agree with averaging for this problem. The parameters ϕ_0, ψ_0 are still free, so for the approximation we have a 2-parameter family of solutions.

9.6.2 Two Coupled Oscillators with Forcing

Consider the system of 2 coupled oscillators with damping and forcing and nonlinear coupling governed by the terms with coefficient $a \neq 0$.

$$\begin{cases} \ddot{x} + x &= \varepsilon(-\mu\dot{x} + axy^2 + h_1 \sin t), \\ \ddot{y} + 2y &+ \varepsilon(-\mu\dot{y} + ax^2 y + h_2 \sin \sqrt{2}t). \end{cases} \tag{9.40}$$

Assume $\mu \geq 0$, h_1, h_2 are constant amplitudes. Consider the equivalent integral equations:

$$\begin{cases} x(t) = r\cos(t + \phi) + \varepsilon \int_0^t \sin(t - s)(-\mu\dot{x}(s) + ax(s)y^2(s) + h_1 \sin s)ds, \\ y(t) = R\cos(\sqrt{2}t + \psi) + \frac{\varepsilon}{\sqrt{2}} \int_0^t \sin(\sqrt{2}t - \sqrt{2}s)(-\mu\dot{y}(s) \\ \quad + ax^2(s)y(s) + h_2 \sin \sqrt{2}s)ds \end{cases} \tag{9.41}$$

As in the preceding example we exclude the normal modes and expand $x(t)$, $y(t)$ and the parameters $r = r_0 + \varepsilon r_1 + O(\varepsilon^2)$, $\phi = \phi_0 + \varepsilon\phi_1 + O(\varepsilon^2)$, $R = R_0 + \varepsilon R_1 + O(\varepsilon^2)$, $\psi = \psi_0 + \varepsilon\psi_1 + O(\varepsilon^2)$. At each order of approximation we have 4 parameter values. Using as lowest order approximation :
$x_0(t) = r_0\cos(t + \phi_0)$, $\dot{x}_0(t) = -r_0\sin(t + \phi_0)$, $y_0(t) = R_0\cos(\sqrt{2}t + \psi_0)$, $\dot{y}_0(t) = -\sqrt{2}R_0\sin(\sqrt{2}t + \psi_0)$ we find for the next step:

$$\tilde{x}(t) = r_0\cos(t + \phi_0) + \varepsilon I_1(t), \quad \tilde{y}(t) = R_0\cos(\sqrt{2}t + \psi_0) + \frac{\varepsilon}{\sqrt{2}} I_2(t)$$

with integral expressions:

$$I_1(t) = \int_0^t \sin(t-s)(\mu r_0 \sin(s+\phi_0) + ar_0 \cos(s+\phi_0)R_0^2 \cos^2(\sqrt{2}s+\psi_0) + h_1 \sin s)ds,$$

and for $I_2(t)$:

$$\int_0^t \sin(\sqrt{2}t - \sqrt{2}s)(\mu\sqrt{2}R_0 \sin(\sqrt{2}s + \psi_0)$$
$$+ ar_0^2 \cos^2(s + \phi_0)R_0 \cos(\sqrt{2}s + \psi_0) + h_2 \sin\sqrt{2}s)ds.$$

Expanding $\sin(t - s) = \sin t \cos s - \cos t \sin s$ and using that r_0, R_0 are positive we find for $I_1(t)$ the quasi-periodic secularity conditions:

$$\mu\sin\phi_0 + \frac{a}{2}R_0^2 \cos\phi_0 = 0, \quad \mu r_0 \cos\phi_0 - \frac{a}{2}r_0 R_0^2 \sin\phi_0 + h_1 = 0. \tag{9.42}$$

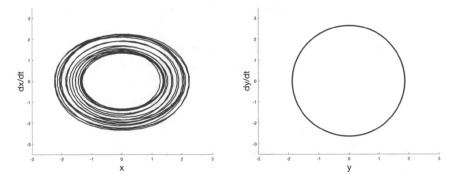

Fig. 9.2 Two projections from 4-space on the $x, \dot{x}$-plane and the $y, \dot{y}$-plane of system (9.40). Corresponding with the quasi-periodic secularity conditions we took $\mu = 1, a = -0.5\sqrt{3}, h_1 = 1.37315, h_2 = -3.26606, \varepsilon = 0.05$ and $x(0) = 0.6866, \dot{x}(0) = -1.1892, y(0) = 1.7321, \dot{y}(0)=1$

Expanding $\sin(\sqrt{2}t - \sqrt{2}s)$ we find for $I_2(t)$ the quasi-periodic secularity conditions:

$$\mu\sqrt{2}\sin\psi_0 + \frac{a}{2}r_0^2\cos\psi_0 = 0, \quad \mu\sqrt{2}R_0\cos\psi_0 - \frac{a}{2}r_0^2 R_0\sin\psi_0 + h_2 = 0. \quad (9.43)$$

We have to exclude $\sin\phi_0 = 0$ and $\sin\psi_0 = 0$ as these values lead to normal mode solutions. Also $\cos\phi_0$ and $\cos\psi_0$ should not vanish. In Fig. 9.2 we show 2 projections of a torus corresponding with parameter values of the quasi-periodic secularity conditions.

We compute the Jacobian J of the 4 secularity conditions to test the uniqueness of the parameter values and the corresponding solutions. We find with abbreviations $\cos\phi_0 = C_f, \sin\phi_0 = S_f, \cos\psi_0 = C_p, \sin\psi_0 = S_p$:

$$J = \begin{pmatrix} 0 & \mu C_f - \frac{a}{2}R_0^2 S_f \\ \mu C_f - \frac{a}{2}R_0^2 S_f & -\mu r_0 S_f - \frac{a}{2}r_0 R_0^2 C_f \\ ar_0 C_p & 0 \\ -ar_0 R_0 S_p & 0 \end{pmatrix}$$

$$\begin{matrix} a R_0 C_f & 0 \\ -ar_0 R_0 S_f & 0 \\ 0 & \mu\sqrt{2}C_p - \frac{a}{2}r_0^2 S_p \\ \mu\sqrt{2}C_p - \frac{a}{2}r_0^2 S_p & -\mu\sqrt{2}R_0 S_p - \frac{a}{2}r_0^2 R_0 C_p \end{matrix} \Bigg) .$$

We use the parameter values of Fig. 9.2. We expect that we can use the implicit function theorem if the rank of the matrix is 4. A MATLAB calculation produces that the determinant of J does not vanish, so the implicit function theorem applies producing for these parameter values a quasi-periodic solution of system (9.40).

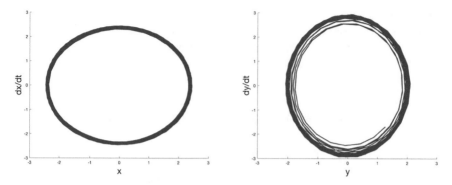

Fig. 9.3 Two projections from 4-space on the $x, \dot{x}$-plane and the $y, \dot{y}$-plane of system (9.40). Corresponding with the quasi-periodic secularity conditions we took $\mu = 1, a = -0.5, h_1 = -2^{1.75}, h_2 = -4, \varepsilon = 0.05$ and $x(0) = 2^{3/4}, \dot{x}(0) = -2^{3/4}, y(0) = \sqrt{2}, \dot{y}(0) = -\sqrt{2}$

An example with different parameter values is shown in Fig. 9.3. The quasi-periodic secularity conditions have been used to produce a torus of system (9.40). In this case a MATLAB calculation of the corresponding Jacobian J shows rank 3, we have a branching producing a 1-parameter family of quasi-periodic solutions at this order of approximation.

9.6.3 Iteration of the Cartoon Problem

The cartoon problem of Sect. 9.4.5 is studied in the context of Neimark-Sacker bifurcation. It is:

$$\ddot{x} + \varepsilon \kappa \dot{x} + (1 + \varepsilon \cos 2t)x + \varepsilon c y = 0,$$
$$\ddot{y} - \varepsilon \mu (1 - y^2)\dot{y} + y + \varepsilon c x = 0.$$

As an alternative approach we will use iteration and the quasi-periodic secularity condition. Averaging produces in a natural way the combination angle $\chi = \phi_1 - \phi_2$. For reasons of comparison we will use transformation (1.6) but change the notation to $r \cos(t + \phi)$ for $x(t)$ and $R \cos(t + \psi)$ for $y(t)$.

The integral equations become:

$$\begin{cases} x(t) & = r \cos(t + \phi) - \varepsilon \int_0^t \sin(t - s)(\kappa \dot{x}(s) + \cos(2s)x(s) + c y(s))ds, \\ y(t) & = R \cos(t + \psi) + \varepsilon \int_0^t \sin(t - s)(\mu(1 - y^2(s))\dot{y}(s) - c x(s))ds. \end{cases}$$

$$(9.44)$$

where we still have to replace x, y etc. by transformation (1.6). As in earlier examples we expand the amplitudes and phases taking into account Remark 9.2. Substitution of the lowest order approximations $x_0(t) = r_0(\cos(t + \phi_o))$, $y_0(t) = R_0 \cos(t + \psi_0)$ produces for $x(t)$ the secularity conditions:

$$
\begin{cases}
-\kappa r_0 \sin \phi_0 + \tfrac{1}{2} r_0 \cos \phi_0 + c R_0 \cos \psi_0 = 0, \\
-\kappa r_0 \cos \phi_0 + \tfrac{1}{2} r_0 \sin \phi_0 - c R_0 \sin \psi_0 = 0.
\end{cases}
\tag{9.45}
$$

Eliminating r_0 / R_0 we find the condition independent of $c \neq 0$:

$$
\frac{\cos \psi_0}{\kappa \, \sin \phi_0 - 0.5 \cos \phi_0} = \frac{-\sin \psi_0}{\kappa \, \cos \phi_0 - 0.5 \sin \phi_0}.
\tag{9.46}
$$

The condition (9.46) for r_0 / R_0 is satisfied if:

$$
(\phi_0, \psi_0) = (\frac{\pi}{4}, -\frac{\pi}{4}), \quad (\phi_0, \psi_0) = (-\frac{\pi}{4}, \frac{\pi}{4}).
\tag{9.47}
$$

We can obtain these values also by requiring system (9.45) to have nontrivial solutions for r_0, R_0. This result agrees with the condition for the angles in Sect. 9.4.5.

The secularity conditions for $y(t)$ become:

$$
\begin{cases}
\mu R_0 \sin \psi_0 (1 - \tfrac{1}{4} R_0^2) + c r_0 \cos \phi_0 = 0, \\
\mu R_0 \cos \psi_0 (1 - \tfrac{1}{4} R_0^2) - c r_0 \sin \phi_0 = 0.
\end{cases}
\tag{9.48}
$$

Choosing the second case of the angles in (9.47) we have:

$$
\kappa r_0 + 0.5 r_0 + c R_0 = 0, \quad \mu R_0 (1 - \frac{1}{4} R_0^2) + c r_0 = 0, \quad \frac{r_0}{R_0} = \frac{-c}{\kappa + 0.5}.
\tag{9.49}
$$

Solving for positive solutions of r_0, R_0 produces a unique periodic solution if the discriminant $\Delta = \det J$ of the Jacobian J does not vanish in (r_0, R_0). We find:

$$
r_0 = \frac{-2c}{\kappa + 0.5} \sqrt{1 - \frac{c^2}{\mu(\kappa + 0.5)}}, \quad R_0 = 2 \sqrt{1 - \frac{c^2}{\mu(\kappa + 0.5)}}.
\tag{9.50}
$$

For the determinant of Jacobian J of the equations for r_0, R_0 in (9.49) we find in this case:

$$
\Delta = (\kappa + 0.5) \mu (1 - \frac{3}{4} R_0^2) - c^2.
\tag{9.51}
$$

For values of the parameters with vanishing Δ a branching of a solution may take place. The condition is not necessary, the local behaviour of the solutions near the parameter values depends on the type of branching phenomenon. For instance a branching from the trivial solutions takes place for the parameter values satisfying (9.51):

$$\kappa = 1.3, \mu = 4, c = 2.6833\ldots$$

For the first case of the angles in Eq. (9.47) we have:

$$-\kappa r_0 + 0.5 r_0 + c R_0 = 0, \quad -\mu R_0 (1 - \frac{1}{4} R_0^2) + c r_0 = 0, \quad \frac{r_0}{R_0} = \frac{-c}{-\kappa + 0.5}, \qquad (9.52)$$

leading to:

$$r_0 = -\frac{2c}{-\kappa + 0.5} \sqrt{1 + \frac{c^2}{\mu(-\kappa + 0.5)}}, \quad R_0 = 2\sqrt{1 + \frac{c^2}{\mu(-\kappa + 0.5)}}. \qquad (9.53)$$

The results agree with the expressions obtained by averaging. For the determinant Δ of the Jacobian J we find:

$$\Delta = (-\kappa + 0.5)\mu(1 - \frac{3}{4} R_0^2) + c^2. \qquad (9.54)$$

Zeros of Δ may correspond with branching phenomena. We have for instance branching from the trivial solutions satisfying Eq. (9.54) if':

$$\kappa = 1, \mu = 4, c = \sqrt{2}.$$

In both cases of the angles the analysis of eigenvalues as in Sect. 9.4.5 provides additional information.

Chapter 10
Averaging for Partial Differential Equations

The qualitative and quantitative theory of weakly nonlinear partial differential equations is still fragmented and it is too early to present a coherent picture of the theory but we will discuss a number of promising methods and interesting problems. For more details and references see the appendix of [58] and [75].

Apart from averaging methods, formal approximation methods, as for example multiple timing, have sometimes been successful and stimulating, both for PDEs on bounded and on unbounded domains. Another formal method that attracted a lot of interest is Whithams approach to combine averaging and variational principles; see for these formal methods [77]. The validation for certain parameter values and side conditions has to be numerical for these methods. One of the reasons of the relatively many open problems for perturbed PDEs is that the unperturbed problem is often already quite difficult and no easy start for a perturbation approach. To study linear perturbation problems can already be very instructive.

The mathematical analysis of asymptotic approximations of PDEs with proofs of validity moves at a slow pace, it rests firmly on the qualitative theory of weakly nonlinear partial differential equations. Existence and uniqueness results are available which involve typically contraction, or other fixed point methods, and maximum principles; one can also use projection methods in Hilbert spaces where eigenfuction expansions are used. If a variational principle is available one can use Galerkin methods, possibly in combination with averaging.

Various forms of averaging techniques are being used in the literature. Some of them are indicated by terms as 'homogenisation' or 'regularisation' methods. Their main purpose is to stabilise numerical integration schemes for partial differential equations. They open the possibility of hybrid methods where asymptotic analysis is used to obtain numerical improvements of a PDE problem.

© The Author(s), under exclusive license to Springer Nature Switzerland AG 2023
F. Verhulst, *A Toolbox of Averaging Theorems*, Surveys and Tutorials in the Applied
Mathematical Sciences 12, https://doi.org/10.1007/978-3-031-34515-9_10

10.1   Metric Spaces and Norms, a Reminder

For metric spaces and norms, also leading to Banach and Hilbert spaces we do not aim at an introduction but we remind the reader only of concepts that are useful in the analysis of PDEs. There are many good introductions to functional analysis directed at applications, see for instance ch. 2 in [39]; the reader is referred to such introductions.

A set of elements $a, b, c, \ldots$ is called a vector space (or linear vectorspace) if in the set multiplication and addition have been defined according to certain rules, for instance with respect to addition the set has to be an Abelian group. The elements can be vectors in the vector space $\mathbb{R}^n$ or for instance the continuous functions or continuously differentiable (C^1) functions on the interval $[0, 1]$.

We can provide the vector space with a norm that means that we assign to each element a of the vector space a unique real number ≥ 0; the norm of a is indicated as $\|a\|$. In $\mathbb{R}^n$ the Euclidean distance for $a = (a_1, a_2, \ldots, a_n)$ (as earlier defined for recurrence) is a norm:

$$\|a\| = \sqrt{a_1^2 + a_2^2 + \ldots + a_n^2}.$$

For ODEs we usually have estimates in the so-called max- or sup-norm. This means that the estimate of an error is given as follows. Consider 2 vector functions $\phi(t), \psi(t)$ on the time-interval $L : 0 \leq t \leq T_0$. With Euclidean norm $\|.\|$ we have sup norm $\|.\|_{sup}$

$$\|\phi - \psi\|_{sup} = \sup_{t \in L} \|\phi(t) - \psi(t)\|.$$

In the case of PDEs one has to estimate functions depending on space and time variables. In Fourier theory one uses the so-called L_2 norm. We put for functions $u : \mathbb{R} \mapsto \mathbb{C}$ the L_2 norm:

$$\|u\|_2 = \left(\int_{\mathbb{R}} u(x)^2 dx \right)^{1/2}.$$

More in general we have for $p = 1, 2, \ldots$ the L_p norm:

$$\|u\|_p = \left(\int_{\mathbb{R}} |u(x)|^p dx \right)^{1/p}.$$

In many problems we consider $x \in D \subset \mathbb{R}^n$, we have to change the domain of integration accordingly.

For normed vector spaces that are Hilbert spaces we can add for each 2 elements f, g the inner product:

$$(f, g) = \int_{\mathbb{R}} f(x)g(x)dx$$

(if the elements are complex functions we replace $g(x)$ by its complex conjugate).

When considering continuous differentiable functions $u(x)$ on a domain D we use the Sobolev inner product :

$$\|u\|_{H^1}^2 = (u, u) + (u_x, u_x).$$

For k times continuously differentiable functions on D we have the Sobolev inner product:

$$\|u\|_{H^k}^2 = (u, u) + (u_x, u_x). + \ldots (u_k, u_k).$$

If $k = 0$ we have the L_2 inner product and norm. Note that estimates in Sobolev norms are for $k \geq 1$ stronger than the L_2 norm as apart from the function values in D it also takes into account its derivatives.

10.2 Averaging a Linear Operator

A typical problem formulation would be to consider the Gauchy problem or an initial-boundary value problem for equations like

$$u_t + Lu = \varepsilon f(u), \quad t > 0, u(0) = u_0. \tag{10.1}$$

L is a linear operator like the Laplace Δ or the wave operator $\Delta - \partial^2/\partial t^2$, $f(u)$ represents the perturbation terms, possibly nonlinear.

To obtain a standard form $u_t = \varepsilon F(t, u)$, suitable for averaging in the case of a PDE can already pose a formidable technical problem, even in the case of simple geometries. However it is reasonable to suppose that one can solve the 'unperturbed' ($\varepsilon = 0$) problem in some explicit form before proceeding to the perturbation problem.

A number of authors, in particular in the former Soviet Union, have addressed problem (10.1). For a survey of results see [48] and [59].

We shall follow the theory developed by Krol in [43] which has some interesting applications on periodic flow in tidal basins. Consider the problem (10.1) with two spatial variables x, y and time t, the coefficients are T-periodic.; we choose $f(u)$ linear; From [43] we have:

Theorem 10.1 *Assume that, after solving the unperturbed problem (10.1), by a variation of constants procedure we can write the problem in the form*

$$\frac{\partial F}{\partial t} = \varepsilon L(t)F, \quad F(x, y, 0) = \gamma(x, y). \tag{10.2}$$

We have on the domain $D \in \mathbb{R}^2$:

$$L(t) = L_2(t) + L_1(t) \tag{10.3}$$

where

$$L_2(t) = b_1(x, y, t)\frac{\partial^2}{\partial x^2} + b_2(x, y, t)\frac{\partial^2}{\partial x \partial y} + b_3(x, y, t)\frac{\partial^2}{\partial y^2}, \tag{10.4}$$

$$L_1(t) = a_1(x, y, t)\frac{\partial}{\partial x} + a_2(x, y, t)\frac{\partial}{\partial y} \tag{10.5}$$

in which $L_2(t)$ is a uniformly elliptic operator on the domain D, L_1, L_2 and so L are T-periodic in t; the coefficients a_i, b_i and γ are C^∞ and bounded with bounded derivatives.

We average the operator L by averaging the coefficients a_i, b_i over t:

$$\bar{a}_i(x, y) = \frac{1}{T}\int_0^T a_i(x, y, t)dt, \quad \bar{b}_i(x, y) = \frac{1}{T}\int_0^T b_i(x, y, t)dt, \tag{10.6}$$

producing the averaged operator $\bar{L}$. As an approximating problem for (10.2) we now take

$$\frac{\partial \bar{F}}{\partial t} = \varepsilon \bar{L}\bar{F}, \quad \bar{F}(x, y, 0) = \gamma(x, y). \tag{10.7}$$

A rather straightforward analysis shows existence and uniqueness of the solutions of problems (10.2) and (10.7) on the time-scale $1/\varepsilon$. We have:

Let F be the solution of initial value problem (10.2) and $\bar{F}$ the solution of initial value problem (10.7), then we have the estimate $\|F - \bar{F}\| = O(\varepsilon)$ on the timescale $1/\varepsilon$.

This is a strong result as the norm $\|.\|$ is the supnorm on the spatial domain and on the time-scale $1/\varepsilon$. By Ben Lemlih and Ellison [12] and, independently in [43] a near-identity transformation is applied to $\bar{F}$ which is autonomous and on which we have explicit information; see for technical details [12].

10.3   Wave Equations, Projection Methods

Consider on a bounded spatial domain hyperbolic, weakly nonlinear equations of the form:

$$u_{tt} + Au = \varepsilon g(u, u_t, t, \varepsilon) \tag{10.8}$$

where A is a positive, selfadjoint linear differential operator on a separable real Hilbert space think of the Laplace operator Δ. Equation (10.8) can be studied by expanding the solutions in eigenfunctions of the operator A but for equivalence of the solutions with Eq. (10.8) this involves solving a system corresponding of an infinite number of ODEs. In many cases, resonance will make this virtually impossible, the averaged (normalised) system is too large, and we have to use truncation techniques; we discuss results from [42] on the asymptotic validity of truncation methods; this yields at the same time interesting information on the timescale of physical interaction of modes.

An example where we can apply expansion in eigenfunctions is the perturbed wave equation

$$u_{tt} - u_{xx} = \varepsilon f(u, u_x, u_t, t, x, \varepsilon), \quad t \geq 0, 0 < x < 1, \tag{10.9}$$

where

$$u(0, t) = u(1, t) = 0, u(x, 0) = \phi(x), u_t(x, 0) = \psi(x), 0 \leq x \leq 1.$$

The eigenfunctions are $\sqrt{2} \sin n\pi x, n = 1, 2, \ldots$ However, a difficulty is that apart from being infinite the system has a resonant spectrum, without the possibility of reduction unless we choose special initial conditions consisting of a number of finite modes. A typical example in the subsequent applications is $f = u^3$; see [77] and Sect. 10.4.4.

An example which is easier to handle is the Klein-Gordon equation with dispersion:

$$u_{tt} - u_{xx} + a^2 u = \varepsilon u^3, \quad t \geq 0, 0 < x < \pi, a > 0. \tag{10.10}$$

We can apply almost-periodic averaging and the averaged system splits into finite-dimensional parts, see the applications. Similar phenomena arise in applications to rod and beam equations.

We will summarise the results for projection methods from [42] and [69].

Consider again the initial-boundary value problem for the nonlinear wave Eq. (10.9). The normalised eigenfunctions of the unperturbed ($\varepsilon = 0$) problem are $v_n(x) = \sqrt{2} \sin(n\pi x), n = 1, 2, \cdots$ and we expand the solution of the initial-boundary value problem for Eq. (10.9) in a Fourier series with respect to these eigenfunctions of the form:

$$u(t, x) = \sum_{n=1}^{\infty} u_n(t) v_n(x). \tag{10.11}$$

As announced above by taking inner products this yields an infinite system of ordinary differential equations which is equivalent to the original problem. The next step is then to truncate this infinite dimensional system (a projection) and apply averaging to the truncated system. The truncation is related to Galerkin's method; this method uses as approximation a finite expansion of the solutions in suitable functions. For our problem one has to *estimate the combined error of truncation and averaging.*

Equation (10.9) with its initial-boundary values has exactly one solution in a suitably chosen Hilbert space $\mathcal{H}_k = H_0^k \times H_0^{k-1}$ where H_0^k are the well-known Sobolev spaces consisting of functions u with derivatives $U^{(k)} \in L_2[0, 1]$ and $u^{(2l)}$ zero on the boundary whenever $2l < k$. Sobolev spaces are natural in physics and engineering as they *measure both positions and derivatives.* It also makes sense to relate the estimate to the energy content of the modes in the system.

It is rather standard to establish existence and uniqueness of solutions *on the timescale* $1/\varepsilon$ under certain mild conditions on f; examples are righthand sides f like u^3, uu_t^2, $\sin u$, $\sinh u_t$ or u^k, $k = 1, 2, \ldots$ etc. Moreover we note that:

1. If $k \geq 3$, u is a classical solution of Eq. (10.9).
2. If $f = f(u)$ is an odd function of u, one can find an even energy integral. If such an integral represents a positive definite energy integral, we are able to prove existence and uniqueness for all time.

In Galerkin's truncation method one considers only N modes of the expansion (10.11) which we shall call the projection u_N of the solution u on a N-dimensional space. To find u_N, we have to solve a $2N$-dimensional system of ordinary differential equations for the expansion coefficients $u_n(t)$ with appropriate (projected) initial values. The estimates for the error $\|u - u_N\|$ depend strongly on the smoothness of the righthandside f of Eq. (10.9) and the initial values $\phi(x)$, $\psi(x)$ but, remarkably enough, not on ε.

Periodic averaging of the truncated system, produces an approximation $\bar{u}_N$ of u_N from the Galerkin-averaging theorem, see Krol [42]. We have:

Theorem 10.2 *Consider the initial-boundary value problem*

$$u_{tt} - u_{xx} = \varepsilon f(u, u_x, u_t, t, x, \varepsilon), \quad t \geq 0, 0 < x < 1$$

where

$$u(0, t) = u(1, t) = 0, u(x, 0) = \phi(x), u_t(x, 0) = \psi(x), 0 \leq x \leq 1.$$

Suppose that f is k-times continuously differentiable and satisfies the existence and uniqueness conditions on the time-scale $1/\varepsilon$, $(\phi, \psi) \in \mathcal{H}_k$; if the solution of the initial-boundary problem is (u, u_t) and the approximationed by the Galerkin-averaging procedure $(\bar{u}_N, \bar{u}_{Nt})$, we have on the timescale $1/\varepsilon$

$$\|u - \bar{u}_N\|_\infty = O(N^{\frac{1}{2}-k}) + O(\varepsilon), \quad N \to \infty, \varepsilon \to 0 \qquad (10.12)$$

$$\|u_t - \bar{u}_{Nt}\|_\infty = O(N^{\frac{3}{2}-k}) + O(\varepsilon), \quad N \to \infty, \varepsilon \to 0. \qquad (10.13)$$

Remark 10.1

- Taking $N = O(\varepsilon^{-\frac{2}{2k-1}})$ we obtain an $O(\varepsilon)$-approximation on the timescale $1/\varepsilon$. So, the required number of modes decreases when the regularity of the data and the order up to which they satisfy the boundary conditions, increases.
- However, this decrease of the number of required modes is not uniform in k. So it is not obvious for which choice of k the estimates are optimal at a given value of ε.
- An interesting case arises if the nonlinearity f satisfies the regularity conditions for all k. This happens for instance if f *is an odd polynomial in u* and with analytic initial values. In such cases the results can be improved by introducing Hilbert spaces of analytic functions (so-called Gevrey classes). The estimates in [42] for the approximations on the time-scale $1/\varepsilon$ obtained by the Galerkin-averaging procedure become in this case

$$\|u - \bar{u}_N\|_\infty = O(N^{-1}a^{-N}) + O(\varepsilon), \quad N \to \infty, \varepsilon \to 0 \qquad (10.14)$$

$$\|u_t - \bar{u}_{Nt}\|_\infty = O(a^{-N}) + O(\varepsilon), \quad N \to \infty, \varepsilon \to 0, \qquad (10.15)$$

where the constant a arises from the bound one has to impose on the size of the strip around the real axis on which analytic continuation is permitted in the initial-boundary value problem.

    The important implication is that, because of the a^{-N}-term we need only $N = O(|log\varepsilon|)$ terms to obtain an $O(\varepsilon)$ approximation on the timescale $1/\varepsilon$.

- It is not difficult to improve the result in the case of finite-modes initial values, i.e. the initial values can be expressed in a finite number of eigenfunctions $v_n(x)$. In this case the error becomes $O(\varepsilon)$ on the timescale $1/\varepsilon$ if N is taken large enough.
- Here and in the sequel we have chosen Dirichlet boundary conditions. It is stressed that this is by way of example and not a restriction. We can also use the method for Neumann conditions, periodic boundary conditions etc.
- It is possible to generalise these results to higher dimensional (spatial) problems; see [42] for remarks and [49] for an analysis of a two-dimensional nonlinear Klein-Gordon equation with Dirichlet boundary conditions on a rectangle. In the case of more than one spatial dimension, many more resonances may be present.
- Related proofs for Galerkin-averaging were given in [25] and [26]. These papers also contain extensions to difference and delay equations.

10.4   Applications

As noted the theory of PDEs is complicated, so to illustrate the ideas our examples will be simplified as much as possible. We will discuss advection-diffusion and nonlinear wave equations, both conservative and dissipative.

10.4.1   Application to a Time-Periodic Advection-Diffusion Problem

As an application one considers in [43] the transport of material (chemicals or sediment) by advection and diffusion in a tidal basin. In this case the advective flow is because of the tides nearly periodic, but diffusive effects are small and take a long time. The inspiration for the model was the transport of chemicals and sediment in the NorthSea, a large tidal basin surrounded by 6 countries, but the formulation applies to similar tidal basins elsewhere.

On a 2-dimensional domain the problem can be formulated as

$$\frac{\partial C}{\partial t} + \nabla.(uC) - \varepsilon \Delta C = 0, \quad C(x, y, 0) = \gamma(x, y), \tag{10.16}$$

where $C(x, y, t)$ is the concentration of the transported material, the flow $u = u_0(x, y, t) + \varepsilon u_1(x, y)$ is given; u_0 is T-periodic in time and represents the tidal flow, εu_1 is a small reststream that is not changing with time. As the diffusion process takes several months close to a year, it is much slower than the tidal period of roughly 12 hours. We are interested in a long timescale approximation. This was also a motivation for the analysis as numerical approximation schemes for this problem turned out to be rather unstable on long timescales.

As usual in these models we assume the flow to be divergence-free; the unperturbed ($\varepsilon = 0$) flow problem is given by

$$\frac{\partial C_0}{\partial t} + u_0 \nabla C_0 = 0, \quad C_0(x, y, 0) = \gamma(x, y), \tag{10.17}$$

a first-order equation which can be integrated along the characteristics with solution of the form $C_0 = \gamma(Q(t)(x, y))$. In the spirit of variation of constants we introduce the change of variables

$$C(x, y, t) = F(Q(t)(x, y), t) \tag{10.18}$$

In actual tidal basins with irregular geometry this is very difficult, see the remark below. We expect F to be slowly time-dependent when introducing (10.18) into the original Eq. (10.16). Using again the technical assumption that the flow $u_0 + \varepsilon u_1$ is divergence-free we find a slowly varying equation of the form (10.2). Note that the assumption of divergence-free flow is not essential, but it simplifies the calculations.

Krol [43] presents some extensions of the theory and explicit examples where the slowly varying equation is averaged to obtain a time-independent parabolic problem.

Remark 10.2 Quite often the latter problem still has to be solved numerically and one may wonder what then the use is of this technique. The answer is, that one needs solutions on a long timescale and that numerical integration of a PDE where the fast periodic oscillations have been eliminated is a much safer procedure.

Remark 10.3 In the analysis presented thus far we have considered unbounded domains. To study the equation on spatially bounded domains, adding boundary conditions, does not present serious obstacles to the techniques and the proofs.

10.4.2 Advection-Diffusion with Reactions and Sources

An extension of the advection-diffusion problem has been obtained in [36]. It is natural to consider reactions of chemicals or sediment using a reaction term f(C). The chemicals tend to get dissolved, the sediments react with the flow. Secondly we include localised sources indicated by B(x,y,t) which in the case of tidal basins can be interpreted as illegal, periodic dumping of chemicals or sediment in the basin. Following [36] the equation becomes

$$\frac{\partial C}{\partial t} + v_0.\nabla C + \varepsilon v_1.\nabla C = \varepsilon \Delta C + \varepsilon f(C) + \varepsilon B(x, y, t), \qquad (10.19)$$

for $t \geq 0$ with initial value $C(x, y, 0) = \gamma(x, y)$. The reaction term $f(C)$ will in general be nonlinear, for instance $f(C) = aC^2$ or $f(C) = aC^5$ depending on the type of chemical reaction. The source term $B(x, y, t$ is periodic in t. Using again variation of constants, we obtain from Eq. (10.19) a perturbation equation in the same way as shown above, but with a more complicated operator $L(t)$. As the tidal period of $v_0(x, y, t)$ is roughly 12 hours, it is natural to assume a common period T with the dumping of chemicals indicated by $B(x, y, t)$. Averaging produces an approximation $\tilde{C}$ of the solution C of the initial value problem (10.19). Interestingly, the result is stronger than in the case without source term B. One can prove that $\tilde{C}$ converges in time to the solution $\tilde{C}_0$ of a time-independent boundary value problem. We have that C converges to a T-periodic solution that is ε-close to $\tilde{C}_0$ for all time. This behaviour is analogous to the case of forced ODEs as we have seen in Chap. 6. The proof is based on maximum principles and the use of suitable sub- and supersolutions; more details are given in [36].

10.4.3   The Wave Equation with Cubic Nonlinearity

This example is inspired by Stroucken and Verhulst [61] and Krol [42]. Consider the equation

$$u_{tt} - u_{xx} = \varepsilon u^3, \quad t \geq 0, 0 < x < \pi, \tag{10.20}$$

with boundary conditions $u(0, t) = u(\pi, t) = 0$ and initial values $u(x, 0) = c_1 \sin x + c_2 \sin 2x$, $u_t(x, 0) = 0, c_1, c_2$ are constants. Using the eigenfunctions of the linearised problem we expand the solution:

$$u(t, x) = \sum_{n=1}^{\infty} u_n(t) \sin nx.$$

According to Remark 10.1 we have to take a number of eigenfunctions that increases if $\varepsilon \to 0$. We restrict ourselves to consider the first 3 modes in the expansion. We find after substitution into Eq. (10.20) and taking inner products for the 3 modes:

$$\begin{cases} \ddot{u}_1 + u_1 &= \frac{3}{4}\varepsilon(u_1^3 + 2u_1 u_2^2 + 2u_1 u_3^2 + u_2^2 u_3 - u_1^2 u_3), \\ \ddot{u}_2 + 4u_2 &= \frac{3}{4}\varepsilon u_2(2u_1^2 + u_2^2 + 2u_3^2 + 2u_1 u_3), \\ \ddot{u}_3 + 9u_3 &= \frac{1}{4}\varepsilon(6u_1^2 u_3 + 6u_2^2 u_3 + 3u_3^3 + 3u_1 u_2^2 - u_1^3). \end{cases} \tag{10.21}$$

The frequencies ω of the linearised system are $1, 2, 3$ so near the origin of phase-space the system is in $1 : 2 : 3$ resonance. As the whole spectrum $\{n^2\}, n = 1, 2, \ldots$ produces resonant frequencies, increasing the number of modes adds to the number of resonances.

System (10.21) has 2 exact normal modes: the u_2 normal mode ($u_1 = u_3 = 0$) and the u_3 normal mode ($u_1 = u_2 = 0$). In addition the system has a 2 dof invariant manifold consisting of the modes u_1, u_3 ($u_2 = \dot{u}_2 = 0$). We discuss approximations of the 3 mode system by introducing slowly varying variables r, ϕ by transformation (1.6) and averaging. As we will not discuss the dynamics of system (10.21) in detail we present only the averaged system for the amplitudes.

$$\begin{cases} \dot{r}_1 &= -\frac{3}{32}\varepsilon r_3 \left(r_2^2 \sin(\phi_1 + \phi_3 - 2\phi_2) - r_1^2 \sin(3\phi_1 - \phi_3)\right), \\ \dot{r}_2 &= 0, \\ \dot{r}_3 &= -\frac{1}{96}\varepsilon r_1 \left(3r_2^2 \sin(\phi_1 + \phi_3 - 2\phi_2) - r_1^2 \sin(\phi_3 - 3\phi_1)\right). \end{cases} \tag{10.22}$$

The first order approximation keeps $r_2(t)$ to $O(\varepsilon)$ constant, a higher order precision calculation is expected to produce nontrivial results. The exchange of energy takes place between the first and third mode. The combination angles are $(\phi_1 + \phi_3 - 2\phi_2)$ and $(\phi_3 - 3\phi_1)$.

Adding a 4th mode u_4 with frequency $\omega_4 = 4$ we will find interaction with mode 2 from terms like $u_1^2 u_4$ in the projection equation for mode 2. This shows again the full resonance character of the problem.

In the notation of [42] we have M from the initial conditions and N for the number of modes. The implication is that with $M = 2, N = 3$ (as above) we have an $O(\varepsilon^{1/4})$ approximation on the timescale $1/\varepsilon^{1/2}$; with $M = 2, N = 4$ we have an $O(\varepsilon^{1/3})$ approximation on the timescale $1/\varepsilon^{5/9}$. A rather weak result.

10.4.4 A 1-Dimensional Dispersive, Cubic Klein-Gordon Equation

As a prototype of a nonlinear wave equation with dispersion consider the nonlinear (cubic) Klein-Gordon equation:

$$u_{tt} - u_{xx} + u = \varepsilon u^3, \quad t \geq 0, 0 < x < \pi \tag{10.23}$$

with boundary conditions $u(0, t) = u(\pi, t) = 0$ and initial values $u(x, 0) = \phi(x), u_t(x, 0) = \psi(x)$ which are supposed to be sufficiently smooth. The problem has been studied by many authors, often by ingenious formal approximation procedures as in [41].

What do we know qualitatively? It follows from the analysis in [42], that we have existence and uniqueness of solutions on the timescale $1/\varepsilon$ and for all time if we put a minus sign on the righthand side. This information has practical use as in some problems, for instance in nonlinear diffusion equations, one may find unbounded solutions.

We start with the eigenfunction expansion (10.11) with:

$$v_n(x) = \sin(nx), \lambda_n^2 = n^2 + 1, n = 1, 2, \cdots$$

for the eigenfunctions and eigenvalues. Substituting this expansion in the Eq. (10.23) and taking the L_2 inner product with $v_n(x)$ for $n = 1, 2, \cdots$ produces an infinite number of coupled ordinary differential equations of the form:

$$\ddot{u}_n + (n^2 + 1)u_n = \varepsilon f_n(u), \quad n = 1, 2, \cdots, \infty$$

with

$$f_n(u) = \sum \sum \sum_{n_1, n_2, n_3 = 1}^{\infty} c_{n_1 n_2 n_3} u_{n_1} u_{n_2} u_{n_3}.$$

As the spectrum is nonresonant (see [61]), we can easily average the complete system or, alternatively, to any truncation number N. The rather trivial result is that the actions are constant to this order of approximation:

$$\dot{r}_i = 0, \ \dot{\phi}_i = \sigma_i(r_1(0), r_2(0), \ldots, r_N(0)), \ i = 1, 2, \ldots, N,$$

with N integrals of motion, the angles are varying slowly as a function of the energy level of the modes.

Considering the theory summarised before, we can make the following observations with regards to the asymptotic character of the estimates:

- In [61] it was proved that, depending on the smoothness of the initial values (ϕ, ψ) we need $N = O(\varepsilon^{-\beta})$ modes (β a positive constant) to obtain an $O(\varepsilon^{\alpha})$ approximation ($0 < \alpha \le 1$) on the timescale $1/\varepsilon$.
- If the initial values can be expressed in a finite number of eigenfunctions $v_n(x)$ (as in the preceding example), it follows from Sect. 10.3, that the error is $O(\varepsilon)$ on the timescale $1/\varepsilon$.
- In [73] a method is developed to prove an $O(\varepsilon)$ approximation on the timescale $1/\varepsilon$ which is applied to the nonlinear Klein-Gordon equation with a quadratic nonlinearity $(-\varepsilon u^2)$.
- As the first order results are not very interesting in [61] a second-order approximation is constructed. It turns out that there exists a small interaction between modes with number n and number $3n$ which probably involves much longer timescales than $1/\varepsilon$. This is still an open problem.
- In [15] one considers the nonlinear Klein-Gordon equation (10.23) in the rather general form

$$u_{tt} - u_{xx} + V(x)u = \varepsilon f(u), \ t \ge 0, 0 < x < \pi \tag{10.24}$$

with V a periodic, even function and $f(u)$ an odd polynomial in u. Assuming rapid decrease of the amplitudes in the eigenfunction expansion (10.11) and diophantine (non-resonance) conditions on the spectrum, it is proved that *infinite-dimensional invariant tori persist in the nonlinear wave Eq. (10.24) corresponding with almost-periodic solutions*. The proof involves a perturbation expansion which is valid on a long timescale.

- In [11] one considers the nonlinear Klein-Gordon equation (10.23) in the more general form

$$u_{tt} - u_{xx} + mu = \varepsilon \phi(x, u), \ t \ge 0, 0 < x < \pi, \tag{10.25}$$

and the same boundary conditions. The function $\phi(x, u)$ is polynomial in u, analytic and periodic in x and odd in the sense that $\phi(x, u) = -\phi(-x, -u)$.

Under a certain non-resonance condition on the spectrum, it is shown in [11] that the solutions remain close to finite-dimensional invariant tori, corresponding with quasi-periodic motion on timescales longer than $1/\varepsilon$.

10.4.5 The Cubic Klein-Gordon Equation on a Square

More spatial dimensions can produce a much richer dynamics. The embedded double resonance, discussed in Chap. 8, arises quite naturally in systems with 2 or more dof; we will discuss a nonlinear wave example developed for 2 spatial dimensions.

Consider the cubic Klein-Gordon equation on a square:

$$u_{tt} - u_{xx} - u_{yy} + u = \varepsilon u^3, \; (x, y) \in [0, \pi] \times [0, \pi], \tag{10.26}$$

with smooth initial conditions. Consider expansion of the solution in a suitable Sobolev space:

$$u(x, y, t) = \sum_{k,l=1}^{\infty} u_{kl}(t) \sin kx \sin ly, \; \omega_{kl}^2 = k^2 + l^2 + 1.$$

One obtains ODE systems by projection on a finite number of modes; the ODEs are of the form:

$$\ddot{u}_{kl} + \omega_{kl}^2 u_{kl} = \varepsilon f_{kl}(u), \; k, l = 1, 2, \ldots$$

As an example we will consider here the 3-mode case $kl = 33, 57, 75$ with frequencies $\omega_{33}^2 = 19, \omega_{57}^2 = \omega_{75}^2 = 75$. This can be handled as a detunend $\frac{1}{2}, 1, 1$ resonance. Other resonance cases are presented in [82]. We find the 3 dof system:

$$\begin{cases} \ddot{u}_{33} + \frac{1}{4}u_{33} = & -\frac{1}{300}u_{33} + \varepsilon\frac{1}{100}(\frac{3}{4}u_{33}^3 + u_{33}u_{57}^2 + u_{33}u_{75}^2), \\ \ddot{u}_{57} + u_{57} = & \varepsilon\frac{1}{100}(\frac{3}{4}u_{57}^3 + u_{33}^2u_{57} + u_{57}u_{75}^2), \\ \ddot{u}_{75} + u_{75} = & \varepsilon\frac{1}{100}(\frac{3}{4}u_{75}^3 + u_{33}^2u_{75} + u_{57}^2u_{75}), \end{cases} \tag{10.27}$$

with $-\frac{1}{300}u_{33}$ as the detuning term. Averaging produces to first order for the amplitudes:

$$\dot{r}_{33} = 0,$$
$$\dot{r}_{57} = -\varepsilon c r_{57} r_{75}^2 \sin 2(\phi_{57} - \phi_{75}),$$
$$\dot{r}_{75} = +\varepsilon c r_{57}^2 r_{75} \sin 2(\phi_{57} - \phi_{75}).$$

The 1 : 1 resonance of u_{57}, u_{75} is effective, but the mode u_{33} is not involved in the first order approximate dynamics, so the dynamics of modes u_{57} and u_{75} dominates to $O(\varepsilon)$ on the timescale $1/\varepsilon$. However, the resonance zones where

$$\sin 2(\phi_{57} - \phi_{75}) = 0,$$

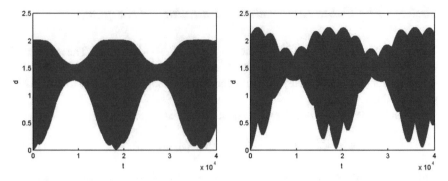

Fig. 10.1 Recurrence from solutions of system (10.27) in 18,000 timesteps for passage through a resonance zone (Euclidean distance d to initial value as a function of time) with $u_{57}(0) = 0.1$, $u_{75}(0) = 1$ and $u_{33}(0) = 0.1, 0.5$. Left a dynamics where the u_{33} mode has little energy. Right a case with considerable energy for the u_{33} mode; in this case we have well-developed tori around the stable periodic solutions causing more obstruction in the passage through resonance process and so more complicated recurrence

present a different picture. In such a resonance zone the modes u_{57} and u_{75} vary very little, we need a second order calculation if we want to take into account quantities that have the same magnitude of variation. Guided by the location of one of the resonance zones we consider a numerical 2-dimensional projection of the amplitudes in a resonance zone; this produces quadratic curves corresponding with the 2 : 4 : 4 resonance. In Fig. 10.1 we show the recurrence when passing a resonance zone by computing the Euclidean distance $d(t)$ to 2 given sets of initial conditions. The 1 : 1 resonance zones of the u_{57} and u_{75} modes are characterised by the first order averaging normal form as:

- Equal amplitudes of modes u_{57}, u_{75};
- Periodic solutions from phase differences $0, \pi/2, \pi, 3\pi/2$;
- Two stable and two unstable periodic solutions on the energy manifold.

The 2 : 4 : 4 resonance is obtained from a higher order calculation (not presented) and is an *embedded double resonance* in these zones.

We observe that a higher order asymptotic calculation produces here new qualitative phenomena.

As we have seen in Chap. 8 the complexity of the resonance zones will generally increase with the dimension. We will find stable and unstable periodic solutions producing homo- and heteroclinic tangles. Solutions that pass through a resonance zone will be delayed by the complexity of the dynamics, but they will always return to the zones, winding for some time around the tori associated with the stable periodic solutions and being perturbed by the tangles. This delay of orbits in resonance zones is called *quasi-trapping* in [84].

Apart from a measure zero number of orbits, passage *through* the resonance zone is the rule and both quasi-trapping and recurrence will take place. By studying the Euclidean deviation (or distance) $d(t)$ from the initial conditions in phase-space as a function of time one may get an indication of the dynamics of the resonance zone inviting further study. Relatively long recurrence times suggest complexity.

10.4.6 The Keller-Kogelman Problem

An interesting example of a nonlinear equation with dispersion and dissipation, generated by a self-excitation (Rayleigh) term, was presented in [40]. Consider the equation

$$u_{tt} - u_{xx} + u = \varepsilon \left(u_t - \frac{1}{3} u_t^3 \right), \quad t \geq 0, 0 < x < \pi, \tag{10.28}$$

with boundary conditions $u(0, t) = u(\pi, t) = 0$ and initial values $u(x, 0) = \phi(x), u_t(x, 0) = \psi(x)$ that are supposed to be sufficiently smooth. As before, putting $\varepsilon = 0$, we have for the eigenfunctions and eigenvalues

$$v_n(x) = \sin(nx), \quad \lambda_n = \omega_n^2 = n^2 + 1, \quad n = 1, 2, \cdots,$$

and again we propose to expand the solution of the initial boundary value problem for the Eq. (10.28) in a Fourier series with respect to these eigenfunctions of the form (10.11). Substituting the expansion into the differential equation we have

$$\sum_{n=1}^{\infty} \ddot{u}_n \sin nx + \sum_{n=1}^{\infty} (n^2 + 1) u_n \sin nx = \varepsilon \sum_{n=1}^{\infty} \dot{u}_n \sin nx - \frac{\varepsilon}{3} (\sum_{n=1}^{\infty} \dot{u}_n \sin nx)^3.$$

When taking inner products we have to Fourier analyse the cubic term. This produces many terms, and it is clear that we will not have exact normal mode solutions, as for instance mode m will excite mode $3m$.

At this point we can start averaging and it becomes important that the spectrum is not resonant. In particular, we have in the averaged equation for u_n only terms arising from $\dot{u}_n^3$ and $\sum_{i \neq n}^{\infty} \dot{u}_i^2 \dot{u}_n$. The other cubic terms do not survive the first order averaging process; the part of the equation for $n = 1, 2, \cdots$ that produces nontrivial terms is:

$$\ddot{u}_n + \omega_n^2 u_n = \varepsilon \left(\dot{u}_n - \frac{1}{4} \dot{u}_n^3 - \frac{1}{2} \sum_{i \neq n}^{\infty} \dot{u}_i^2 \dot{u}_n \right) + \cdots,$$

where the dots stand for nonresonant terms. With the nonresonant terms included this is an infinite system of ordinary differential equations that is still fully equivalent to the original problem.

We can now perform the actual averaging in a notation that contains only minor differences with that of [40]. Transforming in the usual way $u_n(t) = a_n(t) \cos \omega_n t +$

$b_n(t) \sin \omega_n t$, $\dot{u}_n(t) = -\omega_n a_n(t) \sin \omega_n t + \omega_n b_n(t) \cos \omega_n t$, to obtain the standard form, we find after averaging equations for the approximate (a_n, b_n) given by:

$$2\dot{a}_n = \varepsilon a_n \left(1 + \frac{n^2 + 1}{16}(a_n^2 + b_n^2) - \frac{1}{4}\sum_{k=1}^{\infty}(k^2 + 1)(a_k^2 + b_k^2)\right),$$

$$2\dot{b}_n = \varepsilon b_n \left(1 + \frac{n^2 + 1}{16}(a_n^2 + b_n^2) - \frac{1}{4}\sum_{k=1}^{\infty}(k^2 + 1)(a_k^2 + b_k^2)\right).$$

This system shows fairly strong (although not complete) decoupling because of the nonresonant character of the spectrum. Because of the self-excitation, we have no conservation of energy. Putting $a_n^2 + b_n^2 = E_n$, $n = 1, 2, \cdots$, multiplying the first equation with a_n and the second equation with b_n, and adding the equations, we have

$$\dot{E}_n = \varepsilon E_n \left(1 + \frac{n^2 + 1}{16}E_n - \frac{1}{4}\sum_{k=1}^{\infty}(k^2 + 1)E_k\right).$$

We have immediately a nontrivial result: starting in a mode with zero energy, this mode will not be excited on a timescale $1/\varepsilon$. The implication is that if the initial conditions are represented by only N eigenfunctions $u_n(t)v_n(x), n = 1, 2, \ldots$, no other eigenfunctions will be excited on the timescale $1/\varepsilon$.

Another observation is that if we have initially only one nonzero mode, say for $n = m$, the equation for E_m becomes

$$\dot{E}_m = \varepsilon E_m \left(1 - \frac{3}{16}(m^2 + 1)E_m\right).$$

We conclude that with one initial mode we have stable equilibrium at the value

$$E_m = \frac{16}{3(m^2 + 1)}.$$

If we start with initial conditions in a finite number of modes the error is $O(\varepsilon)$, see Sect. 10.3. For related qualitative results see [45].

10.4.7   A Parametrically Excited Linear Wave Equation

Consider the 1-dimensional linear wave equation:

$$u_{tt} - c^2 u_{xx} + \varepsilon \beta u_t + (\omega_0^2 + \varepsilon \phi(t))u = 0, \ t \geq 0, 0 < x < \pi, \tag{10.29}$$

with boundary conditions $u_x(0, t) = u_x(\pi, t) = 0$ and small periodic or almost-periodic excitation $\varepsilon\phi(t)$ with $\omega_0 > 0$ and small damping coefficient $\varepsilon\beta$. In [51] the experimental motivation is discussed; it models for instance a 1-dimensional chain of oscillators with vertical parametric excitation or the linearised behaviour of water waves in a channel with vertical forcing.

Using the eigenfunctions $v_n(x) = \cos nx$ for the Neumann problem (if $\varepsilon = 0$ with eigenvalues $\omega_n^2 = \omega_0^2 + n^2 c^2, n = 0, 1, 2, \ldots$) we expand the solution as:

$$u(x, t) = \sum_0^\infty u_n(t) \cos nx.$$

Substitution in Eq. (10.29) and taking L_2-inner products with $v_n(x)$ produces the system:

$$\ddot{u}_n + \omega_n^2 u_n = -\varepsilon\beta\dot{u}_n - \varepsilon\phi(t)u_n, \; n = 0, 1, 2, \ldots \tag{10.30}$$

with suitable initial conditions. System (10.30) is fully equivalent with Eq. (10.29). The normal mode solutions satisfy the system and correspond with 2-dimensional invariant manifolds.

We conclude that if the term $\phi(t)u_n$ is not resonant, all the modes tend to zero. We consider now a prominent resonance case.

The Mathieu case $\phi(t) = \cos 2t$.

A resonance case occurs of one of the eigenvalues equals 1 or is ε-close to 1. Assume that for certain $m \neq 0$ we have $\omega_m^2 = 1 + \varepsilon\delta$ with δ an $O(1)$ detuning. We find after first order averaging using transformation (1.6) for $n = 0, 1, 2, \ldots$ but $n \neq m$:

$$\dot{r}_n = -\varepsilon\frac{\beta}{2}r_n, \; \dot{\psi}_n = 0. \tag{10.31}$$

If $n \neq m$ the modes are decaying. If $n = m$ we find:

$$\dot{r}_m = -\varepsilon\frac{r_m}{2}(-\beta + \frac{1}{2}\sin 2\psi_m), \; \dot{\psi}_m = \frac{1}{2}\varepsilon(\delta + \frac{1}{2}\cos 2\psi_m). \tag{10.32}$$

If $\beta > 1/2$ the damping exceeds the excitation and the mode u_m decays also. If $0 < \beta < 1/2$ we find 2 solutions with

$$\sin 2\psi_m = 2\beta.$$

They correpond with periodic solutions if simultaneously

$$\delta + \frac{1}{2}\cos 2\psi_m = 0.$$

Elimination of the trigonometric functions produces the relation between β and δ:

$$\beta^2 + \delta^2 = \frac{1}{4},$$

that describe the boundaries of a well-known prominent Floquet instability tongue in parameter space.

In the linear Eq. (10.29) we find that nearly all modes decay but in the case of resonant parametric excitation one mode can be unstable as we know from Floquet theory. This evokes the need for nonlinear modeling and analysis in the applications that follow.

10.4.8   Parametrical Excitation of Nonlinear Waves

In [51] a nonlinear version of a parametrically excited wave equation is studied:

$$u_{tt} - c^2 u_{xx} + \varepsilon\beta u_t + (\omega_0^2 + \varepsilon\phi(t))u = \varepsilon b u^3, \ t \geq 0, 0 < x < \pi, \qquad (10.33)$$

with boundary conditions $u_x(0, t) = u_x(\pi, t) = 0$ and small periodic excitation $\varepsilon\phi(t)$, with $\omega_0 > 0$ and small damping coefficient $\varepsilon\beta$. Additional analysis can be found in [6].

Equation (10.33) models a 1-dimensional nonlinear chain of oscillators with vertical parametric excitation of water waves in a channel with vertical forcing.

Using again the eigenfunctions $v_n(x) = \cos nx$ with eigenvalues $\omega_n^2 = \omega_0^2 + n^2 c^2, n = 0, 1, 2, \ldots$ we expand the solution as:

$$u(x, t) = \sum_0^\infty u_n(t) \cos nx.$$

Substitution in Eq. (10.33) and taking L_2-inner products with $v_n(x)$ produces the system:

$$\ddot{u}_n + \omega_n^2 u_n = -\varepsilon\beta\dot{u}_n - \varepsilon\phi(t)u_n + \varepsilon f_n(\mathbf{u}), \ n = 0, 1, 2, \ldots \qquad (10.34)$$

with $(\mathbf{u}) = (u_0, u_1, u_2, \ldots)$. The components f_n are cubic expressions; the normal mode solutions do not satisfy Eq. (10.34). As an interesting resonance case most attention is given to the choice $\phi(t) = \gamma \cos 2t$. Expanding to 3 modes we have for the cubic terms:

$$\begin{cases} g_0 = u_0^3 + \frac{3}{2}u_0 u_1^2 + \frac{3}{2}u_0 u_2^2 + \frac{3}{4}u_1^2 u_2, \\ g_1 = \frac{3}{4}u_1^3 + 3u_0^2 u_1 + \frac{3}{2}u_1 u_2^2 + 3u_0 u_1 u_2, \\ g_2 = \frac{3}{4}u_2^3 + 3u_0^2 u_2 + \frac{3}{2}u_1^2 u_2 + \frac{3}{2}u_0 u_1^2. \end{cases} \qquad (10.35)$$

The analysis in [51] and [6] deals with many different cases using averaging and numerics. In the numeric analysis 3, 10 and 20 modes are considered. For this complex problem averaging to first and second order produces the skeleton part of the dynamics. Details of the bifurcations and extension for larger values of ε have been studied by numerical exploration guided by the analytic results. We summarise a few aspects.

- If the wave speed c is $O(\varepsilon)$ and ω_0 is an $O(1)$ quantity we have by truncation to a finite number of modes the $1:1:1:\ldots:1$ resonance.
- If the dispersion is small ($\omega_0 = O(\varepsilon)$) system (10.34) is fully resonant. This problem is unsolved.
- The choice $\phi(t) = \gamma \cos 2t$ is considered. If one of the eigenvalues is close to $1/2$ we have the first Floquet resonance. If β is small enough , in fact $\beta < |\gamma|$, one mode will give nontrivial solutions as in the linear case of the preceding subsection, the other modes are not resonant and will decay.
- In [6] the interaction of modes is discussed in more detail. In the case of the combined $1:2$ and $1:1:1$ resonances first and second order averaging is used followed by numerical bifurcations analysis using MATCONT. Averaging to second order is necessary as the terms act for the $1:1:1$ resonance primarily on the cubic terms; the long expressions resulting from second order averaging are presented in an appendix of [6]. One finds from this analysis Hopf bifurcation of limit cycles corresponding with Neimark-Sacker bifurcation to 2-tori and 3-tori; in addition there are many other bifurcations revealing complex dynamics.
- Simulations with more than 3 modes show the presence of more complexity and chaos. In this region second order averaging does not suffice to show all details.

10.4.9 Parametrical Excitation of 2-Dimensional Nonlinear Waves

We will consider the nonlinear wave problem of the preceding subsection on a square domain with special interest in new phenomena. We will find an infinite number of $1:1$ resonances, interesting accidental resonances and complex resonance zones.

Consider the parametrically excited nonlinear wave equation formulated in [51] in the one-dimensional case; we will now consider from [65] the equation on a square.

$$u_{tt} - c^2(u_{xx} + u_{yy}) + \mu u_t + (\omega_0^2 + \beta \cos(\Omega t))u = \alpha u^3, \tag{10.36}$$

where $t \geq 0$ and $0 < x < \pi, 0 < y < \pi$. The Neumann boundary values are $\partial u/\partial n|_S = 0$.

The parameters μ, β are positive and small in a way to be specified.

The system of equations and conditions model the surface deflections $u(x, y, t)$ of a fluid in a square basin with parametric excitation and damping, c is the wave speed.

Resonant nonlinear waves in 2 spatial dimensions were also considered in [49] and [82]. We associate with the system the eigenfunctions:

$$v_{mn}(x, y) = \cos mx \cos ny, m, n = 0, 1, 2 \ldots \tag{10.37}$$

with the space-dependent operator producing the eigenvalues:

$$\omega_{mn}^2 = \omega_0^2 + (m^2 + n^2)c^2, \omega_{mn} = \omega_{nm} = \omega. \tag{10.38}$$

The solutions of Eq. (10.36) with boundary conditions can be approximated by projection to a finite sum of eigenfunctions followed by averaging approximation.

The choice of eigenfunctions is determined by the initial values of Eq. (10.36) while keeping an eye on the resonances of the eigenvalues. It turns out that for the geometry considered here, there are an infinite number of 1 : 1 resonances. This will require our main attention. In addition we will briefly look at prominent accidental resonances.

The Two-Mode 1 : 1 *Resonance*

We propose a two-mode expansion with:

$$u_p(x, y, t) = u_1(t) \cos mx \cos ny + u_2(t) \cos nx \cos my, m, n = 0, 1, 2 \ldots, m \neq n. \tag{10.39}$$

Put $\omega_0 = 1$ and rescale $u = \sqrt{\varepsilon}\bar{u}$ (and its derivatives likewise) in Eq. (10.36) with ε a small positive parameter; we omit the bars. Substituting expansion (10.39) into Eq. (10.36) and taking inner products with the eigenfunctions we find with $\omega_{mn} = \omega, m \neq n$:

$$\begin{cases} \ddot{u}_1 + \omega^2 u_1 = -\mu\dot{u}_1 - \beta u_1 \cos(\Omega t) + \varepsilon\alpha(\frac{9}{16}u_1^3 + \frac{3}{4}u_1 u_2^2), \\ \ddot{u}_2 + \omega^2 u_2 = -\mu\dot{u}_2 - \beta u_2 \cos(\Omega t) + \varepsilon\alpha(\frac{9}{16}u_2^3 + \frac{3}{4}u_1^2 u_2). \end{cases} \tag{10.40}$$

We choose $\Omega = 2\omega$ to study prominent Floquet resonances; rescale $\mu = \varepsilon\bar{\mu}$, $\beta = \varepsilon\bar{\beta}$ after which we omit the bars. System (10.40) contains the 1 : 2 Floquet resonance and in addition the 1 : 1 resonance of the Hamiltonian interaction force.

Note that because of the symmetry of system (10.40) $u_1(t) = \pm u_2(t)$ satisfies the system.

The coordinate planes $u_1, \dot{u}_1$ and $u_2, \dot{u}_2$ are invariant under the phase-flow, we start with the analysis of these *normal mode planes*.

The analysis for both coordinate planes runs exactly along the same lines with symmetric results so we consider only the $u_1, \dot{u}_1$ plane. We put $u_2 = \dot{u}_2 = 0$ and introduce amplitude-phase coordinates as usual by:

$$u_1 = r_1 \cos(\omega t + \psi_1), \dot{u}_1 = -r_1 \omega \sin(\omega t + \psi_1).$$

Deriving the equations for r_1, ψ_1 and averaging over time we find the first order averaged system:

$$\dot{r}_1 = \frac{\varepsilon}{2} r_1 (-\mu + \frac{\beta}{2\omega} \sin 2\psi_1), \quad \dot{\psi}_1 = \frac{\varepsilon}{4\omega} (\beta \cos 2\psi_1 - \alpha \frac{27}{32} r_1^2). \qquad (10.41)$$

A critical point corresponding with an equilibrium of system (10.41) is given by:

$$\beta \sin 2\psi_1 = 2\mu\omega, \quad \beta \cos 2\psi_1 = \alpha \frac{27}{32} r_1^2, \quad 0 < \frac{2\mu\omega}{\beta} < 1. \qquad (10.42)$$

We can eliminate the phase angle to find:

$$r_1^2 = r_0^2 = \frac{32}{27\alpha} \sqrt{\beta^2 - 4\mu^2 \omega^2}. \qquad (10.43)$$

Computing eigenvalues at the critical point shows that the periodic solution is stable within the invariant coordinate plane. For the eigenvalues we have:

$$\lambda_{1,2} = -\mu \pm \sqrt{5\mu^2 - \frac{\beta^2}{\omega^2}}. \qquad (10.44)$$

If $\beta > \sqrt{5}\mu\omega$ the periodic solution is complex stable in the coordinate plane, if $2\mu\omega < \beta < \sqrt{5}\mu\omega$ the periodic solution is stable with real eigenvalues. If $\beta = 2\mu\omega$ the periodic solution vanishes.

An important question is whether the periodic solution is stable or unstable in the full 4-dimensional system. For $u_2, \dot{u}_2$ near zero we should not use polar coordinates. Instead we introduce with (1.16) in system (10.40) the slowly-varying variables a, b by:

$$u_2 = a \cos \omega t + \frac{b}{\omega} \sin \omega t, \quad \dot{u}_2 = -a\omega \sin \omega t + b \cos \omega t.$$

Introducing amplitude-phase variables by (1.6) for u_1 and a, b variables for u_2 in system (10.40) we have to average the system. To determine the stability of the normal mode periodic solution we compute the matrix of coefficients of the averaged system for r_1, ψ_1, a, b and find the eigenvalues of its gradient, the Jacobian at the periodic $u_1(t)$ for $a = b = 0$. This means that we can leave out the quadratic and cubic expressions in a, b. For the averaged system in the variables r_1, ψ_1, a, b

we find:

$$
\begin{cases}
\dot{r}_1 &= \frac{\varepsilon}{2} r_1 (-\mu + \frac{\beta}{2\omega} \sin 2\psi_1) + \dots, \\
\dot{\psi}_1 &= \frac{\varepsilon}{2} (\frac{\beta}{2\omega} \cos 2\psi_1 - \alpha \frac{27}{64\omega} r_1^2) + \dots, \\
\dot{a} &= \frac{\varepsilon}{2} (-\mu a + \frac{\beta}{2\omega^2} b + \alpha \frac{3}{16\omega} r_1^2 (a \sin 2\psi_1 + 2\frac{b}{\omega} - \frac{b}{\omega} \cos 2\psi_1)) + \dots, \\
\dot{b} &= \frac{\varepsilon}{2} (-\mu b - \frac{\beta}{2} a + \alpha \frac{3}{16} r_1^2 (2a + a \cos 2\psi_1 - \frac{b}{\omega} \sin 2\psi_1)) + \dots.
\end{cases}
\tag{10.45}
$$

where the dots stand for the omitted higher order terms in a, b. The Jacobian at the periodic solution in the coordinate plane becomes when omitting the factor $\varepsilon/2$:

$$
\begin{pmatrix}
0 & \frac{r_0}{\omega} \beta \cos 2\psi_1 & 0 & 0 \\
-\frac{27\alpha}{32\omega} r_0 & -2\mu & 0 & 0 \\
0 & 0 & -\mu + \frac{3\alpha}{16\omega} r_0^2 \sin 2\psi_1 & \frac{\beta}{2\omega^2} + \frac{3\alpha}{16\omega^2} r_0^2 (2 - \cos 2\psi_1) \\
0 & 0 & -\frac{\beta}{2} + \frac{3\alpha}{16} r_0^2 (2 + \cos 2\psi_1) & -\mu - \frac{3\alpha}{16\omega} r_0^2 \sin 2\psi_1
\end{pmatrix}.
\tag{10.46}
$$

The 4 eigenvalues are splitting up in 2 groups; the first two correspond with the eigenvalues of Eq. (10.44), the second group produces the eigenvalues $\lambda_{3,4}$ with $\lambda_3 + \lambda_4 = -2\mu$. We find:

$$
\lambda_{3,4} = -\mu \pm \sqrt{ \frac{13}{108} \frac{\beta^2}{\omega^2} - \frac{88}{81} \mu^2 - \frac{128}{81} \frac{\omega^2 \mu^4}{\beta^2} }.
\tag{10.47}
$$

$\lambda_{3,4}$ depends on the parameters μ, β, ω, α. We conclude that the 2 periodic normal mode solutions of the 1 : 1 resonances are asymptotically stable if $(39/2)^{1/4} \beta \le \omega\mu$. See Fig. 10.2.

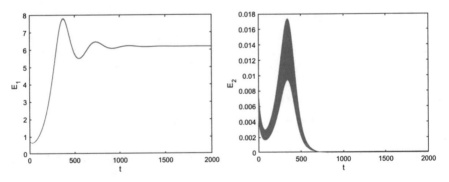

Fig. 10.2 The behaviour of the solutions of system (10.40) near the invariant u_1, $\dot{u}_1$ coordinate plane is shown in [65] by plotting $E_1(t) = 0.5(\dot{u}_1^2(t) + 6u_1^2(t))$ and $E_2(t) = 0.5(\dot{u}_2^2(t) + 6u_2^2(t))$ for the parametrically excited oscillators. The initial conditions are $u_1(0) = 0.5$, $\dot{u}_1(0) = 0$, $u_2(0) = \dot{u}_2(0) = 0.05$; $\omega^2 = 6$, $\mu = 0.01$, $\beta = 0.1$, $\alpha = 0.05$

In [65] a detailed analysis is given for general position orbits of the $1 : 1$ resonance with criteria for the existence and stability of periodic solutions. Both in the dissipative case $\mu > 0$ and the volume-preserving case $\mu = 0$ the general position periodic solutions are unstable suggesting a selection in the wave equation to normal mode solutions.

Additional results are given in [65] for coupled $1 : 1$ resonant systems and cases like the $1 : 1 : 3$ and $1 : 1 : 1$ resonances.

A natural question is how the results obtained for the nonlinear wave equation with parametric excitation (10.36) in [65] is affected by the geometry of the domain. The square domain produces resonances like $1 : 1$ because of its symmetry. One can choose for instance a rectangle with unequal sides like $0 < x < \pi, 0 < y < L\pi, ,$ $L \neq 1$. The eigenfunctions become in the case of Neumann conditions:

$$v_{mn}(x, y) = \cos mx \cos \frac{n}{L} y, m, n = 0, 1, 2 \ldots$$

with eigenvalues of the space-dependent operator producing:

$$\omega_{mn}^2 = \omega_0^2 + \left(m^2 + \frac{n^2}{L^2} \right) c^2.$$

A two-mode expansion with:

$$u_p(x, y, t) = u_1(t) \cos m_1 x \cos \frac{n_1}{L} y + u_2(t) \cos m_2 x \cos \frac{n_2}{L} y,$$

gives 2 perturbed equations like system (10.40) but with 3 different frequencies $\omega_1, \omega_2, \Omega$. In general there will be no resonance or a resonance will take place between 2 frequencies. There are many interesting cases to be analysed for this problem.

How to perform the analysis for a more general bounded 2-dimensional domain remains an open question.

References

1. C. Abdulwahed, F. Verhulst, Recurrent canards producing relaxation oscillations. CHAOS **31**(2) (2021). https://doi.org/10.1063/5.0040726
2. G.M. Andersen, J.F. Geer, Power series expansions for the frequency and period of the limit cycle of the van der Pol equation. SIAM J. Appl. Math. **42**, 678–693 (1982).
3. V.I. Arnold, *Geometric Methods in the Theory of Ordinary Differential Equations* (Springer, BErlin, 1983). Original Russian ed. 1978
4. V.I. Arnold, V.V. Kozlov, A.I. Neishtadt, Mathematical aspects of classical and celestial mechanics, in *Dynamical Systems*, vol. III, ed. by V.I. Arnold (Springer, Berlin, 1988)
5. T. Bakri, R. Nabergoj, A. Tondl, F. Verhulst, Parametric excitation in nonlinear dynamics. Int. J. Nonlinear Mech. **39**, 311–329 (2004).
6. T. Bakri, H.G.E. Meijer, F. Verhulst, Emergence and bifurcations of Lyapunov manifolds in nonlinear wave equations. J. Nonlinear Sci **19**, 571–596 (2009). https://doi.org/10.1007/s00332-009--9045-2
7. T. Bakri, F. Verhulst, Bifurcations and quasi-periodic dynamics: torus breakdown. Z. Angew. Math. Phys. **65**, 1053–1076 (2014)
8. T. Bakri, Y.A. Kuznetsov, F. Verhulst, Torus bifurcations in a mechanical system. J. Dyn. Differ. Equ. **27**, 371–403 (2015)
9. T. Bakri, F. Verhulst, Time-reversal, tori families and canards in the Sprott A and NE9 systems. CHAOS (2022). https://doi.org/10.1063/5.0097508
10. T. Bakri, F, Verhulst, From A. Tondl's Dutch contacts to Neimark-Sacker-bifurcation. Appl. Comput. Mech. **16** (2022). https://doi.org/10.24132/acm.2022.770
11. D. Bambusi, On long time stability in Hamiltonian perturbations of nonresonant linear pde's. Nonlinearity **12**, 823–850 (1999)
12. A. Ben Lemlih, J.A. Ellison, Method of averaging and the quantum anharmonic oscillator. Phys. Rev. Lett. **55**, 1950–1953 (1985)
13. N.N. Bogoliubov, Y.A. Mitropolsky, *Asymptotic Methods in the Theory of Nonlinear Oscillations* (Gordon and Breach, Washington, 1961)
14. H. Bohr, *Almost-Periodic Functions* (Chelsea, 1947). Reprint
15. J. Bourgain, Construction of approximative and almost periodic solutions of perturbed linear Schrödinger and wave equations. GAFA **6**, 201–230 (1996)
16. H.W. Broer, G.H. Huitema, M.B. Seyvruk, *Quasi-Periodic Motions in Families of Dynamical Systems*. Lecture Notes in Mathematics, vol. 1645 (Springer, Berlin, 1996)
17. H.W. Broer, H.M. Osinga, G. Vegter, Algorithms for computing normally hyperbolic invariant manifolds. ZAMP **48**, 480–524 (1997)

© The Author(s), under exclusive license to Springer Nature Switzerland AG 2023
F. Verhulst, *A Toolbox of Averaging Theorems*, Surveys and Tutorials in the Applied Mathematical Sciences 12, https://doi.org/10.1007/978-3-031-34515-9

18. H.W. Broer, M.B. Sevryuk, KAM theory: quasi-periodicity in dynamical systems, in *Handbook of Dynamical Systems* vol. 3, Chap.6, ed. by H.W, Broer, B. Hasselblatt, F. Takens (Elsevier, Amsterdam, 2010), pp. 251–344

19. S.-N. Chow, J.K. Hale, *Methods of Bifurcation Theory* (Springer, Berlin, 1982)

20. E.A. Coddington, N. Levinson, *Theory of Ordinary Differential Equations* (McGraw-Hill Book, New York, 1955)

21. J.J. Duistermaat, Non-integrability of the 1:1:2-*resonance*. Ergodic Theory Dynamical Syst. **4**, 553–568 (1984)

22. R.M. Evan Iwanowski, *Resonance Oscillations in Mechanical Systems* (Elsevier, Amsterdam, 1976)

23. S. Fatimah, M. Ruijgrok, Bifurcation in an autoparametric system in 1:1 internal resonance with parametric excitation. Int. J. Non-Linear Mech. **37**, 297—308 (2002)

24. P. Fatou, *Sur le mouvement d'un système soumis á des forces á courte période*. Bull. Soc. Math. **56**, 98–139 (1928)

25. M. Fečkan, A Galerkin-averaging method for weakly nonlinear equations. Nonlinear Anal. **41**, 345–369 (2000)

26. M. Fečkan, Galerkin-averaging method in infinite-dimensional spaces for weakly nonlinear problems, in *Progress in Nonlinear Differential Equations and Their Applications*, vol. 43, ed. by H.R. Grosinho, M. Ramos, C. Rebelo, L. Sanches (Birkhäuser Verlag, Basel, 2001)

27. J. Ford, The Fermi-Pasta-Ulam problem: paradox turns discovery. Phys. Rep. **213**, 271–310 (1992)

28. M. Golubitsky, I. Stewart, *The Symmetry Perspective* (Birkhäuser Verlag, Basel, 2000)

29. G. Gorelik, A. Vitt, *Oscillations of an elastic pendulum as an example for two parametrically excited vibratory sysstems* (Russ.). J. Tech. Phys. USSR col. 3 (1933)

30. J. Grasman, Asymptotic methods of relaxation oscillations and applications. Appl. Math. Sci. **63**, (1987)

31. J. Guckenheimer, P. Holmes, Nonlinear oscillations, dynamical systems and bifurcations of vector fields. Appl. Math. Sci. **42**, 5th printing (1996)

32. R. Haberman, Energy bounds for the slow capture by a center in sustained resonance. SIAM J. Appl. Math. **43**, 244–256 (1983)

33. J.K. Hale, *Oscillations in Nonlinear Systems* (MacGraw-Hill, New York, 1963). Repr. Dover Publ., New York, 1992

34. J.K. Hale, *Ordinary Differential Equations* (Wiley-Interscience, New York, 1969)

35. M. Henon, C. Heiles, The applicability of the third integral oi motion: some numerical experiments. Astron. J. **69**, 73–79 (1964)

36. J.J. Heijnekamp, M.S. Krol, F. Verhulst, Averaging in non-linear transport problems. Math. Meth. Appl. Sci. **18**, 437–448 (1995)

37. M. Hirsch, C. Pugh, M. Shub, *Invariant Manifolds*. Lecture Notes in Mathematics, vol. 583 (Springer, Berlin, 1977)

38. S. Jafari, J.C. Sprott, S. Golpayegani, Elementary quadratic chaotic flows with no equilibria. Phys. Lett. A **377**, 699–702 (2013)

39. T. Kapitula, K. Promislow, Spectral and dynamical stability of nonlinear waves. Appl. Math. Sci. **185** (2013)

40. J.B. Keller, S. Kogelman, Asymptotic solutions of initial value problems for nonlinear partial differential equations. SIAM J. Appl. Math. **18**, 748–758 (1970)

41. J. Kevorkian, J.D. Cole Perturbation methods in applied mathematics. Appl. Math. Sci. **34** (1981)

42. M.S. Krol, On a Galerkin-averaging method for weakly non-linear wave equations. Math. Meth. Appl. Sci. **11**, 649–664 (1989)

43. M.S. Krol, On the averaging method in nearly time-periodic advection-diffusion problems. SIAM J. Appl. Math. **51**, 1622–1637 (1991)

44. C. Kuehn, Multiple time scale dynamics. Appl Math. Sci. **191** (2015)

45. J. Kurzweil, Van der Pol perturbation of the equation for a vibrating string. Czech. Math. J. **17**, 558–608 (1967)

46. Y.A. Kuznetsov, *Elements of Applied Bifurcation Theory*, Rev. edn. (Springer, Berlin, 2004)
47. Matcont, Numerical continuation and bifurcation program. Available at http://www.matcont. ugent.be
48. Y.A. Mitropolsky, G. Khoma, M. Gromyak, *Asymptotic Methods for Investigating Quasiwave Equations of Hyperbolic Type* (Kluwer Academic Publishers, New York, 1997).
49. H. Pals, The Galerkin-averaging method for the Klein-Gordon equation in two space dimensions. Nonlinear Anal. **27**, 841–856 (1996)
50. Henri Poincaré, *Les Méthodes Nouvelles de la Mécanique Céleste*, 3 vols (Gauthier-Villars, Paris, 1892/1893/1899).
51. R.H. Rand, W.I. Newman, B.C. Denardo, A.I. Newman, Dynamics of a nonlinear parametrically-excited partial differential equation, in Proceeding of the Design Eng. Techn. Conferences, vol. 3 (1995), pp. 57–68. ASME, DE-84-1 (see also Newman et al. Chaos 9, pp. 242–253 (1999))
52. B. Rink, F. Verhulst, *Near-integrability of periodic FPU-chains*. Phys. A **285**, 467–482 (2000)
53. B. Rink, Symmetry and resonance in periodic FPU-chains. Commun. Math. Phys. **218**, 665–685 (2001)
54. M. Roseau *Vibrations Nonlinéaires et Théorie de la Stabilité* (Springer, Berlin/Heidelberg, 1966)
55. D. Ruelle, F. Takens On the nature of turbulence. Commun. Math. Phys. **20**, 167–192 (1971)
56. J.A. Sanders, Are higher order resonances really interesting? Celestial Mech. **16**, 421–440 (1977)
57. J.A. Sanders, F. Verhulsts, Approximations of higher order resonances with an application to Contopoulos' model problem. Asymptotic Anal. LNM **711**, 209–228 (1979)
58. J.A. Sanders, F. Verhulst, and J. Murdock, *Averaging Methods in Nonlinear Dynamical Systems*, Rev edn. (Springer-Verlag, New York, 2007)
59. A.L. Shtaras, The averaging method for weakly nonlinear operator equations. Math. USSSR Sbornik **62**, 223–242 (1989)
60. J.J, Stoker, *Nonlinear Vibrations in Mechanical and Electrical Systems* (1950). Repr. as Wiley Classics Library Edition 1992
61. A.C.J. Stroucken, Verhulst, The Galerkin-averaging method for nonlinear, undamped continous systems. Math. Meth. Appl. Sci. **9**, 520–549 (1987)
62. A. Tondl, *On the Interaction Between Self-Excited and Parametric Vibrations*. Monographs and Memoranda vol. 25 (National Research Institute for Machine Design, Bechovice, Prague, 1978)
63. A. Tondl, M. Ruijgrok, F. Verhulst, R. Nabergoj, *Autoparametric Resonance in Mechanical Systems* (Cambridge University Press, Cambridge, 2000), 196 pp.
64. J.M. Tuwankotta, F. Verhulst , Symmetry and resonance in Hamiltonian systems. SIAM J. Appl. Math. **61**, 1369–1385 (2000)
65. F. Verhulst, J.M. Tuwankotta, A parametrically excited nonlinear wave equation, in *Nonlinear Dynamics of Discrete and Continuous Systems*, ed. by A.K. Abramian et al. Advanced Structural Materials, vol. 139 (2020). https://doi.org/10.1007/978-3-030-53006-8-11
66. B. Van den Broek, F. Verhulst, Averaging techniques and the oscillator flywheel problem. Nieuw Arch. Wiskunde 4th series, **5**, 185–106 (1987)
67. B. Van den Broek, *Studies in Nonlinear Resonance, Applications of Averaging*. Ph.D Thesis, University of Utrecht, 1988
68. E. Van der Aa, First-order resonances in three-degrees-of-freedom systems. Celestial Mech. **31**, 163–191 (1983)
69. E. Van der Aa, M.S. Krol, *A Weakly Nonlinear Wave Equation with Many Resonances*. Ph.D Thesis, M.S. Krol, University of Utrecht, 1990
70. A.H.P. Van der Burgh, On the asymptotic solutions of the differential equations of the elastic pendulum. J. de Mécanique **4**, 507–520 (1968)
71. A.H.P. Van der Burgh, *Studies in the Asymptotic Theory of Nonlinear Resonance*. PhD Thesis, Technical University Delft, 1974

72. A. H. P. Van der Burgh, On the asymptotic approximations of the solutions of a system of two non-linearly coupled harmonic oscillators. J. Sound Vibr. **49**, 93–103 (1976)

73. W.T. Van Horssen, *Asymptotics for a class of semilinear hyperbolic equations with an application to a problem with a quadratic nonlinearity*. Nonlinear Anal. TMA **19**, 501–530 (1992)

74. F. Verhulst, *Discrete symmetric dynamical systems at the main resonances with applications to axi-symmetric galaxies*. Philos. Trans. R. Soc. Lond. **290**, 435–465 (1979)

75. F. Verhulst, On averaging methods for partial differential equations, in *SPT98-Symmetry and Perturbation Theory II*, ed. by A. Degasperis, G. Gaeta (World Scientific, Singapore, 1999), pp. 79–95

76. F. Verhulst, *Nonlinear Ordinary Differential Equations and Dynamical Systems*, 2nd Rev. edn. (Springer, Berlin, 2000)

77. F. Verhulst, *Methods and Applications of Singular Perturbations, Boundary Layers and Timescale Dynamics*. Texts in Applied Mathematics, vol. 50 (Springer, Berlin, 2005)

78. F. Verhulst, *Henri Poincaré, Impatient Genius* (Springer, Berlin, 2012)

79. F. Verhulst, Profits and pitfalls of timescales in asymptotics. SIAM Rev. **57**, 255–274 (2015)

80. F. Verhulst, Near-integrability and recurrence in FPU cell-chains. Int. J. Bifurcation Chaos **26**(14) (2016). https://doi.org/10.1142/S0218127416502308.

81. F. Verhulst, Interaction of low and higher order Hamiltonian resonances. Int. J. Bifurcation Chaos **28**(08) (2018). https://doi.org/10.1142/S0218127418500979

82. F. Verhulst, Recurrence and resonance in the cubic Klein-Gordon equation. Acta Appl. Math. **162**, 145–164 (2019). https://doi.org/10.1007/s10440-019-00238-4

83. F. Verhulst, Evolution to mirror-symmetry in rotating systems. Symmetry (MDPI) **13**, 1189–1208 (2021)

84. G.M. Zaslavsky, *The Physics of Chaos in Hamiltonian Systems*, 2nd ext. edn. (Imperial College Press, London, 2015)

Index

© The Author(s), under exclusive license to Springer Nature Switzerland AG 2023

F. Verhulst, *A Toolbox of Averaging Theorems*, Surveys and Tutorials in the Applied
Mathematical Sciences 12, https://doi.org/10.1007/978-3-031-34515-9

Printed in the United States
by Baker & Taylor Publisher Services